Lecture Notes in Physics

Edited by H. Araki, Kyoto, J. Ehlers, München, K. Hepp, Zürich
R. Kippenhahn, München, D. Ruelle, Bures-sur-Yvette
H. A. Weidenmüller, Heidelberg, J. Wess, Karlsruhe and J. Zittartz, Köln

Managing Editor: W. Beiglböck

356

R. Wehrse (Ed.)

Accuracy of Element Abundances from Stellar Atmospheres

Proceedings of Two Sessions
Allocated at the IAU General Assembly
in Baltimore, USA, August 1988

Springer-Verlag
Berlin Heidelberg GmbH

Editor

Rainer Wehrse
Institut für Theoretische Astrophysik
Im Neuenheimer Feld 561, 6900 Heidelberg, FRG

ISBN 978-3-662-13781-9   ISBN 978-3-540-46971-1 (eBook)
DOI 10.1007/978-3-540-46971-1

© Springer-Verlag Berlin Heidelberg 1990
Originally published by Springer-Verlag Berlin Heidelberg New York in 1990
Softcover reprint of the hardcover 1st edition 1990

2153/3140-543210 – Printed on acid-free paper

Preface

A classical stellar atmosphere is fully characterized by the effective temperature, the gravity, the geometrical extension, the turbulent velocity and the abundances of the chemical elements and their isotopes. It is evident that the abundance pattern of a stellar atmosphere contains most of the information available about the star's structure and its past and future evolution (although many species are formed together and therefore their abundances are not independent) In addition, stellar abundances provide important constraints on the evolution of the cluster or galaxy the star is in.

This information can be obtained essentially only by spectral analysis, which, however, comprises many simplifying assumptions and a very large number of laboratory input data. In addition, uncertainties are introduced by the absolute calibration of the spectra as well as their limited wavelength resolution and their finite signal-to-noise ratio. The resulting inaccuracies of the element abundances cannot be determined in a general way since the corresponding expressions are complex matrix equations and depend on the element(s) under consideration as well as the spectral type.

In order to discuss the present situation of abundance accuracies for main sequence stars two sessions were allocated at the IAU General Assembly in Baltimore (August 1988). After a general introduction, in the first part the accuracies of atomic transition probabilities, ionisation and excitation cross-sections and of line broadening data were discussed. In the second part the specific problems and achievements for different stellar types were presented, and finally an overview on the possibilities of abundance determinations for stars outside the Galaxy was given. The concentration on main sequence stars seemed to be necessary since otherwise a large variety of types (e.g. cool and hot giants, white dwarfs, Wolf-Rayet stars, magnetic stars) would have each required an additional contribution and would have made the meeting very inhomogeneous since a complete coverage would have been impossible. This volume presents somewhat enlarged versions of the talks given in Baltimore.

It is a pleasure for me to thank the President of IAU Commission 36, Prof. Dr. K. Kodaira, for the original proposal for the sessions. Many discussions with Prof. B. Baschek helped me to find a suitable framework. I thank Dr. D. Innes for critically reading many manuscripts and I am indebted to Mrs. M. Wolf for typing several contributions.

Heidelberg   Rainer Wehrse
November, 1989

CONTENTS

INTRODUCTION

RAINER WEHRSE
Institut für Theoretische Astrophysik
Im Neuenheimer Feld 561
D 6900 Heidelberg, FRG

The accurate determination of element abundances from a stellar atmosphere is a complicated process. Ideally, it involves the fit of a synthetic spectrum to the observed one for the whole wavelength range where there is some radiative flux. The abundances of all elements should then be determined consistently with the basic parameters effective temperature and gravity (and possibly also photospheric radius as well as micro- and macroturbulence). In practice, fortunately, most elements do not influence the pressure and temperature distribution of the atmosphere and show up only in a more or less large number of isolated spectral lines. Therefore the abundances of such elements may be determined with fixed temperature and pressure stratifications and independently from those of other species without significant loss in the accuracy.

In order to assess the errors in abundances derived from the spectral analysis of stars we first have to realize that there are many uncertainties, which have consequences for the abundances, both in the observed spectra as well as in the models. Roughly these uncertainties can be grouped into the following categories:
 (i) Inadequacies in the physical assumptions entering the model construction (e.g. the neglect of mechanical energy flux, vertical and horizontal inhomogeneities);
 (ii) Inaccuracies in the mathematics of the model construction (discretisation of the governing differential equations, convergence);
 (iii) Uncertainties in the atomic physics data entering the model (e.g. absorption cross-sections, transition probabilities);
 (iv) Uncertainties in the observed fluxes (which may be due to
    a) a limited signal-to-noise ratio,
    b) uncertainties in the calibration of the spectrograph and the detector system,
    c) the limited spectral resolution of the observations (for a review see Cayrel, 1988).

The consequences of the uncertainties in the physical assumptions are very difficult to estimate, since a systematic sensitivity analysis is not possible. Therefore, the only way "to obtain a feeling" for these abundance errors seems to be the comparison of results based on different sets of assumptions and their consistency with all available observations (cf. Gustafsson, 1988).

Such an empirical analysis is also frequently performed to evaluate the errors resulting from (ii) to (iv); see e.g. Cayrel de Strobel (1985) and subsequent contributions in this volume. However, as will be shown below, the errors can also be evaluated in a more formal way.

Since in most cases the interdependences between the atmospheric parameters are not clear a priori let us assume the general case that all are closely connected. Therefore, we consider the (very long) super-vector X, which contains the parameter vector p (with the components effective temperature T_{eff}, gravity g, microturbulence v_{turb}), the abundances of all M (~ 92) elements (considered as a vector ϵ with the components ϵ_i), and the relevant data of the input physics (a vector $\mathbf{q}$ containing as elements wavelengths of the edges in the continuum, transition probabilities, cross-sections, etc.), i.e.

$$\mathbf{X} = (\mathbf{p}, \epsilon, \mathbf{q})^t. \tag{1}$$

The corresponding error vector is

$$\Delta\mathbf{X} = (\Delta\mathbf{p}, \Delta\epsilon, \Delta\mathbf{q})^t. \tag{2}$$

$\mathbf{X}$ determines the fluxes F_λ at all wavelengths λ via
  (i) the hydrostatic equation;
 (ii) the radiative transfer equations for all λ;
(iii) the energy equation (in most cases the equation of radiative + convective equilibrium) for all depths;
(iv) the equations describing the thermodynamical and optical properties of the matter as e.g. the equation of state and the expressions for the absorption and scattering coefficients as functions of the local temperature and pressure.

We can therefore formally write for the emergent fluxes at wavelength λ

$$F_\lambda = F_\lambda(\mathbf{X}) \tag{3}$$

and for the errors

$$\Delta F_\lambda = \frac{dF_\lambda}{d\mathbf{X}}\Delta\mathbf{X}. \tag{4}$$

It is seen that for the evaluation of the uncertainties in the abundances we have to know the errors $\Delta\mathbf{p}$ and $\Delta\mathbf{q}$ and the uncertainties in the fluxes at M wavelengths (subsequently designated by the vector $\Delta\mathbf{F}_\lambda$), which are e.g. given by $\Delta F/(S/n)$ in the case that the signal-to-noise ratio is much smaller than other error sources for the flux. Therefore we can write for the maximal error

$$\frac{d\mathbf{F}_\lambda}{d\epsilon}\Delta\epsilon = \Delta\mathbf{F}_\lambda + \frac{d\mathbf{F}_\lambda}{d\mathbf{p}}\Delta\mathbf{p} + \frac{d\mathbf{F}_\lambda}{d\mathbf{q}}\Delta\mathbf{q} \tag{5}$$

or

$$\Delta\epsilon = \left(\frac{d\mathbf{F}_\lambda}{d\epsilon}\right)^{-1}\left(\Delta\mathbf{F}_\lambda + \frac{d\mathbf{F}_\lambda}{d\mathbf{p}}\Delta\mathbf{p} + \frac{d\mathbf{F}_\lambda}{d\mathbf{q}}\Delta\mathbf{q}\right). \tag{6}$$

In the Appendix it will be shown that the Jacobi matrices which indicate the strength of the dependences can be evaluated analytically and a full sensitivity analysis can be performed. However, the expressions are very complicated and depend on all specific model parameters as well as on the chosen wavelength set so that no numerical data will be given here. In addition, only after the specification of all details can it be derived which of the three terms on the RHS of (6) is the most important one. On the other hand, even without the knowledge of the matrices it is seen that $\Delta\epsilon$ does not depend on the absolute values of the fluxes, since a common factor (which e.g. can be due to an incorrect distance) cancels.

From (6) it may seem advantageous to choose wavelengths where the fluxes are particularly sensitive to the element abundances, i.e. $(d\mathbf{F}_\lambda/d\epsilon)$ is diagonally dominated and the norm of the inverse is small. However, in such cases the fluxes are often dominated by a single process and the last term in the RHS sum of (6) may be large so that $\Delta\epsilon$ may still be rather large.

It is also evident that the use of many more wavelength points than elements can improve the accuracy of the composition as long as the errors of the input data are not systematic. In such a case the errors in the abundances have to be found from (6) by a least-squares approach.

Appendix

The matrices of (6) can all be evaluated analytically as we will show here by the example of $dF/d\epsilon$. For this purpose we consider a stellar atmosphere consisting of N shells and describe the radiative transfer by means of transmission and reflection matrices $\mathbf{t}_n^\pm$ and $\mathbf{r}_n^\pm$ as well as source vectors $\mathbf{q}_n^\pm B$ (cf. Peraiah, 1984; Schmidt and Wehrse, 1987), i.e. for the intensities $\mathbf{I}_n^+$ of photons leaving shell n in the outward direction, $\mathbf{I}_{n-1}^-$ of photons leaving the shell n in the inward direction and the corresponding intensities $\mathbf{I}_n^-$ and $\mathbf{I}_{n-1}^+$ for photons entering the shell we have the relation

$$\begin{pmatrix} \mathbf{I}_n^+ \\ \mathbf{I}_{n-1}^- \end{pmatrix} = \begin{pmatrix} \mathbf{t}_n^+ & \mathbf{r}_n^- \\ \mathbf{r}_n^+ & \mathbf{t}_n^- \end{pmatrix} \begin{pmatrix} \mathbf{I}_{n-1}^+ \\ \mathbf{I}_n^- \end{pmatrix} + \begin{pmatrix} \mathbf{q}_n^+ & 0 \\ 0 & \mathbf{q}_n^- \end{pmatrix} \begin{pmatrix} \tilde{\mathbf{b}} \\ \tilde{\mathbf{b}} \end{pmatrix}, \tag{A1}$$

where the first term on the RHS describes the transmission and reflection and the second one the internal sources. The matrices have to be determined from the radiative transfer equation which we assume to have the general form

$$\frac{d}{dz} \begin{pmatrix} \mathbf{I}^+ \\ \mathbf{I}^- \end{pmatrix} = \begin{pmatrix} \alpha & \beta \\ \gamma & \delta \end{pmatrix} \begin{pmatrix} \mathbf{I}^+ \\ \mathbf{I}^- \end{pmatrix} + \begin{pmatrix} \mathbf{b} \\ \mathbf{b} \end{pmatrix}. \tag{A2}$$

Equation (A1) couples the radiation fields of shells $(n\text{-}1)$, n, and $(n\text{+}1)$ so that the combined equations for all layers can now be conveniently written in matrix notation:

$$\mathbf{AI} = \mathbf{QB}, \tag{A3}$$

i.e. the intensities are given by

$$\mathbf{I} = \mathbf{A}^{-1}\mathbf{QB}, \tag{A4}$$

and the flux can now be expressed by

$$\mathbf{F} = \mathbf{w}\mathbf{A}^{-1}\mathbf{QB}, \tag{A5}$$

where the matrix $\mathbf{w}$ contains the weights for the integration over angle. Note that $\mathbf{F}$ describes the depth dependence of the flux at a given wavelength; the vector $\mathbf{F}_\lambda$ used above consists of the the first elements of vectors $\mathbf{F}$ for different wavelengths.

Assuming radiative equilibrium we can now write the total flux which is directly related to the effective temperature T_{eff}

$$\mathbf{F}_{\text{tot}} = \sigma T_{\text{eff}}^4 = \sum_i a_i \mathbf{w} \mathbf{A}_i^{-1} \mathbf{Q}_i \mathbf{B}_i. \tag{A6}$$

The coefficients a_i are the weights for the integration over wavelength. The derivatives of the fluxes with respect to the abundances ϵ_i can now be written

$$\frac{d\mathbf{F}}{d\epsilon} = \frac{\partial \mathbf{F}}{\partial \epsilon} + \frac{\partial \mathbf{F}}{\partial \mathbf{T}} \frac{\partial \mathbf{T}}{\partial \epsilon} + \frac{\partial \mathbf{F}}{\partial \mathbf{P}} \frac{\partial \mathbf{P}}{\partial \epsilon}. \tag{A7}$$

The first term on the RHS gives the direct changes in the flux for given temperature and pressure stratifications, the second and the third ones describe the indirect changes via modifications in the temperatures and' pressures. The tensors of the RHS of (A7) are given by

$$\frac{\partial \mathbf{F}}{\partial \epsilon} = \mathbf{w} \left\{ \frac{\partial \mathbf{A}^{-1}}{\partial \epsilon} \mathbf{QB} + \mathbf{A}^{-1} \frac{\partial \mathbf{Q}}{\partial \epsilon} \mathbf{B} \right\}$$

$$= \mathbf{w} \left\{ \mathbf{A}^{-1} \frac{\partial \mathbf{Q}}{\partial \epsilon} \mathbf{B} - \mathbf{A}^{-1} \frac{\partial \mathbf{A}}{\partial \epsilon} \mathbf{A}^{-1} \mathbf{QB} \right\}$$

$$= \mathbf{w} \mathbf{A}^{-1} \left\{ \frac{\partial \mathbf{Q}}{\partial \epsilon} \mathbf{B} - \frac{\partial \mathbf{A}}{\partial \epsilon} \mathbf{I} \right\}, \tag{A8}$$

$$\frac{\partial \mathbf{F}}{\partial \mathbf{T}} = \mathbf{w} \mathbf{A}^{-1} \left\{ \frac{\partial \mathbf{Q}}{\partial \mathbf{T}} \mathbf{B} - \frac{\partial \mathbf{A}}{\partial \mathbf{T}} \mathbf{I} + \mathbf{Q} \frac{\partial \mathbf{B}}{\partial \mathbf{T}} \right\}, \tag{A9}$$

and

$$\frac{\partial \mathbf{T}}{\partial \epsilon} = -\frac{\partial \{\sigma \mathbf{T}_{eff}^4 - \sum_i a_i \mathbf{w} \mathbf{A}_i^{-1} \mathbf{Q}_i \mathbf{B}_i\}}{\partial \epsilon} \left[\frac{\partial \{\sigma \mathbf{T}_{eff}^4 - \sum_i a_i \mathbf{w} \mathbf{A}_i^{-1} \mathbf{Q}_i \mathbf{B}_i\}}{\partial \mathbf{T}} \right]^{-1}$$

$$= -\sum_i a_i \mathbf{w} \mathbf{A}_i^{-1} \left\{ \frac{\partial \mathbf{Q}_i}{\partial \epsilon} \mathbf{B}_i - \frac{\partial \mathbf{A}_i}{\partial \epsilon} \mathbf{I}_i \right\}$$

$$\times \left[\sum_i a_i \mathbf{w} \mathbf{A}_i^{-1} \left\{ \frac{\partial \mathbf{Q}_i}{\partial \mathbf{T}} \mathbf{B}_i - \frac{\partial \mathbf{A}_i}{\partial \mathbf{T}} \mathbf{I}_i + \mathbf{Q}_i \frac{\partial \mathbf{B}_i}{\partial \mathbf{T}} \right\} \right]^{-1}. \tag{A10}$$

The last term in (A7) can be evaluated in an analogous way.

Since the matrices $\mathbf{Q}_i$ and $\mathbf{A}_i$ consist of transmission and reflection matrices (in addition to some geometry factors, see Wehrse, 1981) and the derivatives of these matrices are known (cf. Schmidt and Wehrse, 1987) it is also straightforward to build up the tensors $\partial \mathbf{Q}_i / \partial \epsilon$, $\partial \mathbf{Q}_i / \partial \mathbf{T}$, $\partial \mathbf{A}_i / \partial \epsilon$, and $\partial \mathbf{A}_i / \partial \mathbf{T}$ from

$$\frac{\partial t^+}{\partial \epsilon} = \frac{\partial t^+}{\partial \tau} \frac{\partial \tau}{\partial \epsilon} = (-t^+ \alpha - t^+ \beta r^-) \frac{\partial \tau}{\partial \epsilon}, \tag{A11}$$

$$\frac{\partial t^-}{\partial \epsilon} = \frac{\partial t^-}{\partial \tau} \frac{\partial \tau}{\partial \epsilon} = (\delta t^- - r^- \beta t^-) \frac{\partial \tau}{\partial \epsilon}, \tag{A12}$$

$$\frac{\partial r^+}{\partial \epsilon} = \frac{\partial r^+}{\partial \tau} \frac{\partial \tau}{\partial \epsilon} = (-t^+ \beta t^-) \frac{\partial \tau}{\partial \epsilon}, \tag{A13}$$

$$\frac{\partial r^-}{\partial \epsilon} = \frac{\partial r^-}{\partial \tau} \frac{\partial \tau}{\partial \epsilon} = (\gamma + \delta r^- - r^- \beta r^- - r^- \alpha) \frac{\partial \tau}{\partial \epsilon}, \tag{A14}$$

$$\frac{\partial t^+}{\partial T} = \frac{\partial t^+}{\partial \tau}\frac{\partial \tau}{\partial T} = \left(-t^+\alpha - t^+\beta r^-\right)\frac{\partial \tau}{\partial T}, \tag{A15}$$

$$\frac{\partial t^-}{\partial T} = \frac{\partial t^-}{\partial \tau}\frac{\partial \tau}{\partial T} = \left(\delta t^- - r^-\beta t^-\right)\frac{\partial \tau}{\partial T}, \tag{A16}$$

$$\frac{\partial r^+}{\partial T} = \frac{\partial r^+}{\partial \tau}\frac{\partial \tau}{\partial T} = \left(-t^+\beta t^-\right)\frac{\partial \tau}{\partial T}, \tag{A17}$$

$$\frac{\partial r^-}{\partial T} = \frac{\partial r^-}{\partial \tau}\frac{\partial \tau}{\partial T} = \left(\gamma + \delta r^- - r^-\beta r^- - r^-\alpha\right)\frac{\partial \tau}{\partial T}. \tag{A18}$$

Finally, the τ-derivatives are directly derived from the definition of the optical depth:

$$\frac{\partial \tau}{\partial \epsilon} = \sum_{i,j}\int_0^z \sigma_{i,j}\frac{dn_{i,j}}{d\epsilon}dz, \tag{A19}$$

$$\frac{\partial \tau}{\partial T} = \sum_{i,j}\int_0^z \frac{d(\sigma_{i,j}n_{i,j})}{dT}dz, \tag{A20}$$

where $\sigma_{i,j}$ indicates the absorption cross-section of species i being in the internal state j and $n_{i,j}$ refers to the corresponding number density.

References:

Cayrel, R.: 1988, Proc. IAU Symp. 132, 345

Cayrel de Strobel, G.: 1985, Proc. IAU Symp. 111, 137

Gustafsson, B.: 1988, Proc. IAU Symp. 132, 333

Peraiah, A.: 1984, Methods in radiative transfer, W. Kalkofen, ed., Cambridge University Press, Cambridge, 281

Schmidt, M., Wehrse, R.: 1987, Numerical radiative transfer, W. Kalkofen, ed., Cambridge University Press, Cambridge, 341

Wehrse, R.: 1981, Monthly Not. Royal Astron. Soc. 195, 553

ON THE ACCURACY OF ATOMIC TRANSITION PROBABILITIES

W. L. Wiese and J. R. Fuhr
National Institute of Standards and Technology
Gaithersburg, MD 20899, USA

A. Introduction

Spectral lines are principally characterized by three quantities: their wavelengths
and associated spectroscopic classifications; the energies of their upper and lower
atomic levels; and their oscillator strengths or atomic transition probabilities. For
spectral lines formed in dense gases or plasmas one must add as additional quantities
the line shape and line shift parameters. Of the three principal quantities, atomic
transition probabilities are much less known than either wavelengths or energy levels.
Even for the small percentage of lines for which they are reliably known, the accuracy
of the available transition probability data is usually not very high and a number of
discrepancies between various data are encountered. Thus, a review of the status of
the accuracy of transition probability data and of recent progress seems to be
appropriate.

B. Data Availability

Before the status of accuracies is discussed it is appropriate to take a look at the
availability of reliable transition probability data.[1-5] To do this, we utilize the
comprehensive bibliographical material collected and catalogued at the Data Center on
Atomic Transition Probabilities at the National Institute of Standards and Technology
(formerly the National Bureau of Standards). We also make use of the critical
tabulations, issued by this data center, which are generally limited to data which are
estimated to be accurate within ±50%. Figure 1 shows an overview of the data
availability for the various stages of ionization of all chemical elements. In this
arrangement, the chemical elements are listed in order of increasing atomic number Z
versus the stage of ionization, given by Roman numerals. Numeral I indicates the
neutral spectrum, II the singly ionized spectrum, etc. In this graph, isoelectronic
sequences appear as diagonal lines. Spectra, for which an adequate number of reliable
data are available, i.e., with uncertainties estimated to be smaller than ±50%, are
shown as black squares; spectra for which only few data are available are denoted by an
"X." To be in this category, we require that at least 10 lines originating in more
than two transition arrays are available for simple spectra, and about 50 lines from
more than three arrays are available for complex spectra. All other spectra--with
fewer or no data available--are left blank. The figure shows that the data situation
is satisfactory only for the very light elements and becomes rather poor as the atomic
number increases beyond iron (Z = 26). Data for heavier elements are not yet available
for many stages of ionization except for many neutral and singly ionized spectra and a
number of highly stripped ions belonging to isoelectronic sequences of relatively
simple outer atomic structure, such as alkali-like or alkaline earth-like ions or

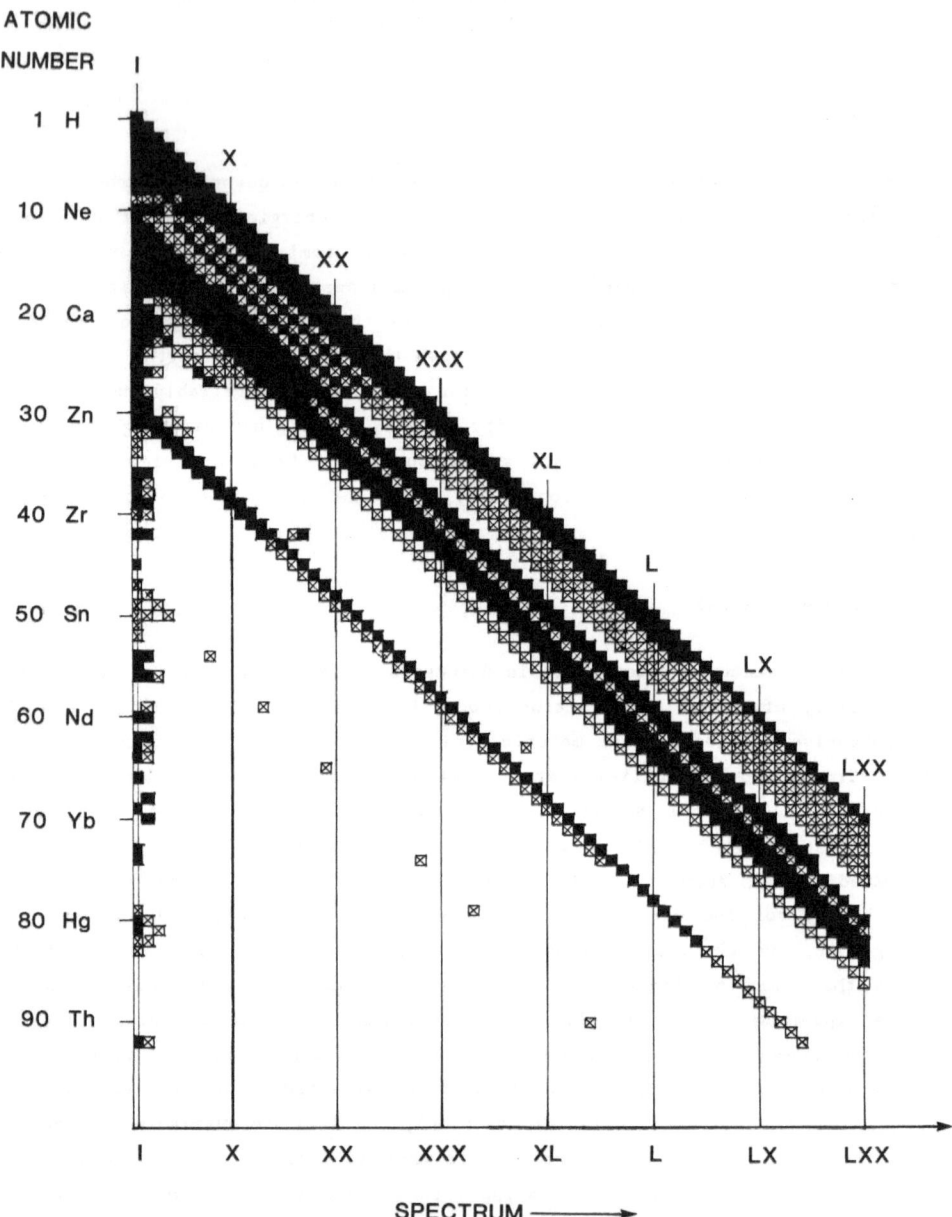

Fig. 1.   Availability of reliable atomic transition probability data. Spectra with an adequate amount of data are shown as black squares; spectra with some reliable data are denoted by an "X"; all other spectra--with very few or no data available--are left blank (see text for details).

ions or sequences which are isoelectronic with the lighter elements (H, He, Li, Be, B, Al, Si, P, S). For the spectra of several neutral and singly ionized heavy atoms, only relatively few reliable data exist and are thus shown as "X's." This is in contrast to the black squares for most spectra of the lighter elements where reliable transition probability data--many estimated to be in the 10-25% accuracy range--are available for a fairly large fraction of the prominent spectral lines. It should be noted that the availability of data for highly stripped ions is almost entirely due to advances in atomic structure theory. For isoelectronic ions, the transition probability of some fixed transition can be readily calculated as a function of nuclear charge. Good progress has been achieved in recent years on the major calculational problem for highly stripped ions--the inclusion of relativistic effects, so that reliable data may be calculated up to very highly ionized atoms (see, e.g., Ref. 6).

C. Summary of Principal Techniques:

Before we discuss the accuracy of atomic transition probabilities, a summary of the major methods[7] to determine these quantities is in order. Experimentally, the majority of the data has been determined by the emission technique. In this technique the data are derived from measurements of line intensities which are proportional to the product of population of the radiating atomic level, photon energy and transition probability. A major problem is the determination of the level population, for which usually the model of local thermodynamic equilibrium (LTE) is invoked and measurements of the temperature and a particle density are required.

A technique quite similar in spirit is the absorption method, which has been very successfully applied to the spectra of many transition metals. In this technique, the data are also obtained from a line intensity measurements, but these are now the intensities of absorption lines from continuous background emission. Again, a critical problem is the determination of the populations of the absorbing atomic levels. Another, currently less frequently applied method is the measurement of anomalous dispersion in the vicinity of a spectral line, utilizing the "hook"-technique. In this technique, the proportionality between the strength of a spectral line and the anomalous change in the index of refraction across a spectral line is utilized to determine the former.

Another very important and successful technique has been the measurement of lifetimes of excited atomic states. This method yields radiative decay times, which are inversely proportional to the sum of all transition probabilities arising from a given upper level. For many prominent lines, such as the principal resonance lines, the sum reduces essentially to a single term for the only possible or the dominant transition.

In recent years the combination of a few very accurate lifetime measurements--to provide absolute data--with measurements of numerous transitions in emission or absorption on a relative scale has been used to great advantage, for two reasons: first, relative emission or absorption measurements may be efficiently and accurately accomplished (much more so than the same measurements on an absolute scale); secondly, a single lifetime measurement--which can be done quite accurately--is sufficient to put the relative data on a reliable absolute scale, provided that the complete transition probability sum from a given atomic level has been determined.

On the theoretical side, two principal approaches are utilized. On one hand, _ab initio_ atomic structure calculations are widely used, which are normally based on the self-consistent field (Hartree-Fock) method. The accuracy of this method has been significantly advanced by multiconfiguration treatments to account for electron correlation. Entirely different in character are semi-empirical approaches, which recently have been mostly used on complex heavy atoms and ions. This approach requires the use of experimentally available energy-level data.

D. Critical Assessment of Data

Atomic transition probabilities and the related atomic lifetimes, as obtained from the above-mentioned methods, have been critically evaluated and compiled by us in the NIST data center on atomic transition probabilities for more than 25 years. The following five criteria are being used in the assessment process:[2,3]

1. We base each assessment on a general evaluation of the applied method, which has gradually evolved over the years.

2. We examine the authors' consideration of "critical factors", that are encountered in each experimental or theoretical approach.

3. We assess the authors' uncertainty estimate, especially with respect to completeness.

4. We check the extent of agreement of the results with other reliable data by utilizing graphical or tabular comparisons.

5. We test the data for their degree of fit into systematic trends or regularities.

Of probably highest importance is the second criterion, involving the "critical factors" of each method. All of these factors must have been considered in order for systematic uncertainties to be minimized. To explain what specifically is meant by the critical factors,[3] these are listed below for one of the major experimental methods, the emission technique. For this technique, four critical factors have been found to be of principal importance:

a. Accurate diagnostics of the emission source, i.e. reliable determinations of particle densities and temperature.

b. Validity of the plasma model utilized for the emission source. Usually the local thermodynamic equilibrium (LTE) model is used and optically thin conditions for the line emission are assumed. The validity of these assumptions may be tested with theoretical validity criteria or experimental techniques (especially the optical thickness).

c. A well-defined emission source, i.e., homogeneity in the line of sight, temporal and spatial stability and small boundary layers.

d. Accurate line intensity measurements that include contributions from extended line wings and are based on accurate radiometric standards.

E. Some Specific Examples:

The spectrum of neutral iron is of great astrophysical interest and, mainly because of this importance, much experimental work has been performed on this spectrum. Among the most significant pieces of work have been the very accurate absorption measurements by Blackwell and co-workers,[8-12] and the comprehensive emission experiment by Bridges and Kornblith[13] with a wall-stabilized arc. Blackwell and co-workers claim that typical uncertainties of their data are of order ±5% on the absolute scale while Bridges and Kornblith estimate uncertainties of about ±25% or smaller for their absolute data. In both experiments the absolute scale has been determined from the average of recent lifetime results for the resonance transition. Nearly the same set of lifetime data was chosen by the two groups; therefore, the two absolute scales agree closely with each other. It is instructive to compare these two sets of data graphically. For this purpose, we have made plots of the transition probability ratios on a logarithmic scale. The data are expressed as log gf, where g is the statistical weight of the lower level of the transition, and f is the oscillator strength. This f-value, derived from classical electron theory, is equivalent to the

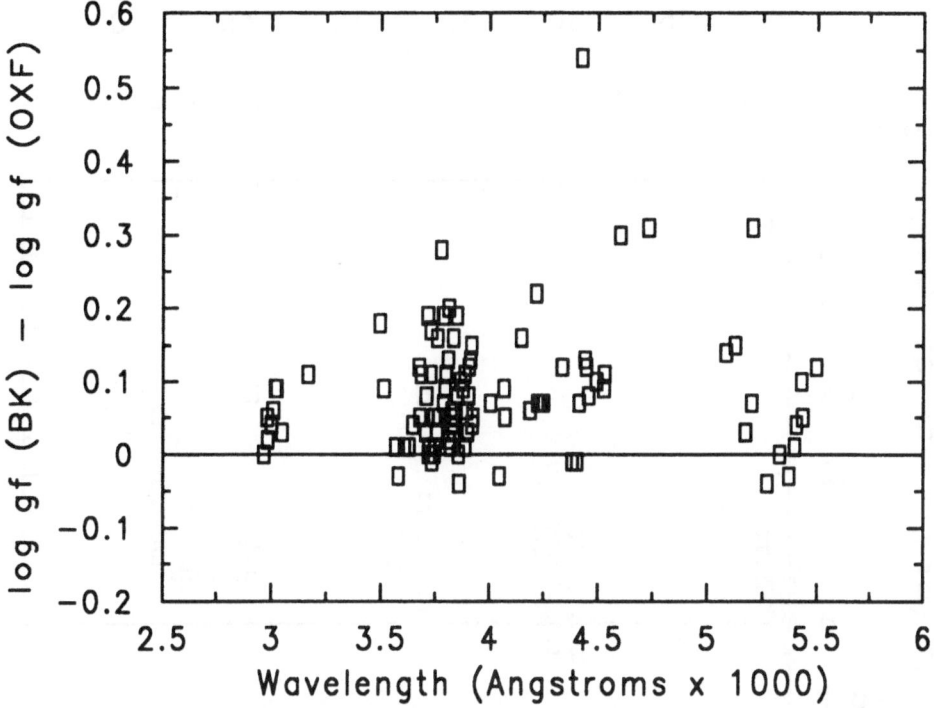

Fig. 2. Logarithmic ratios of the gf-values of Bridges and Kornblith (emission technique) and Blackwell, et al. (the "Oxford group", absorption technique) versus wavelength (in Å) of the transitions. The transition yielding a very high value for the ratio is probably affected by a blend.

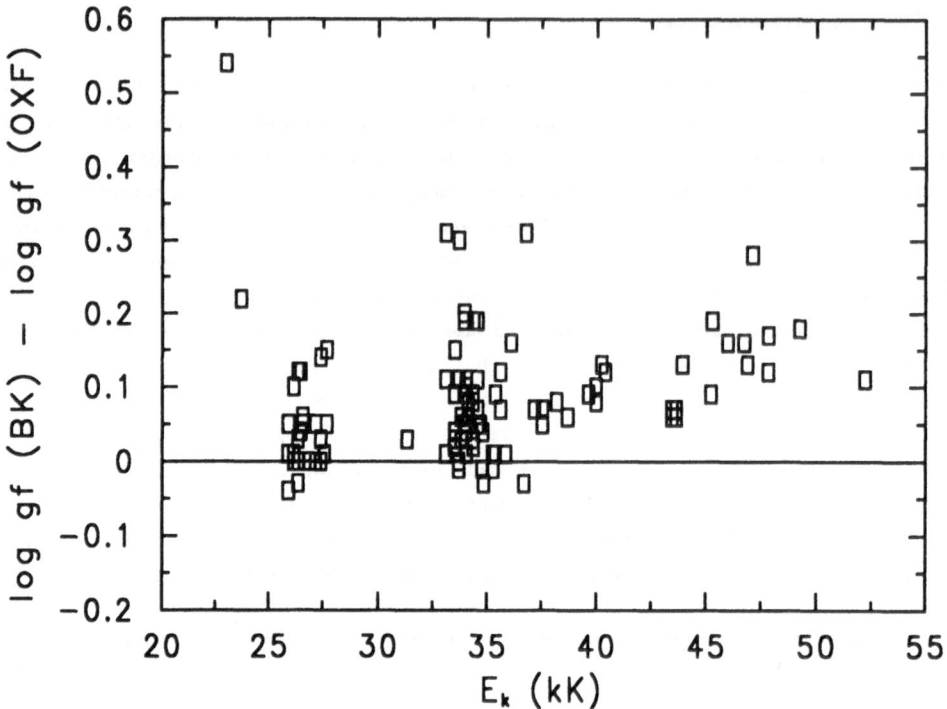

Fig. 3. Same ratios as in Fig. 2, but plotted versus upper energy level of the transitions.

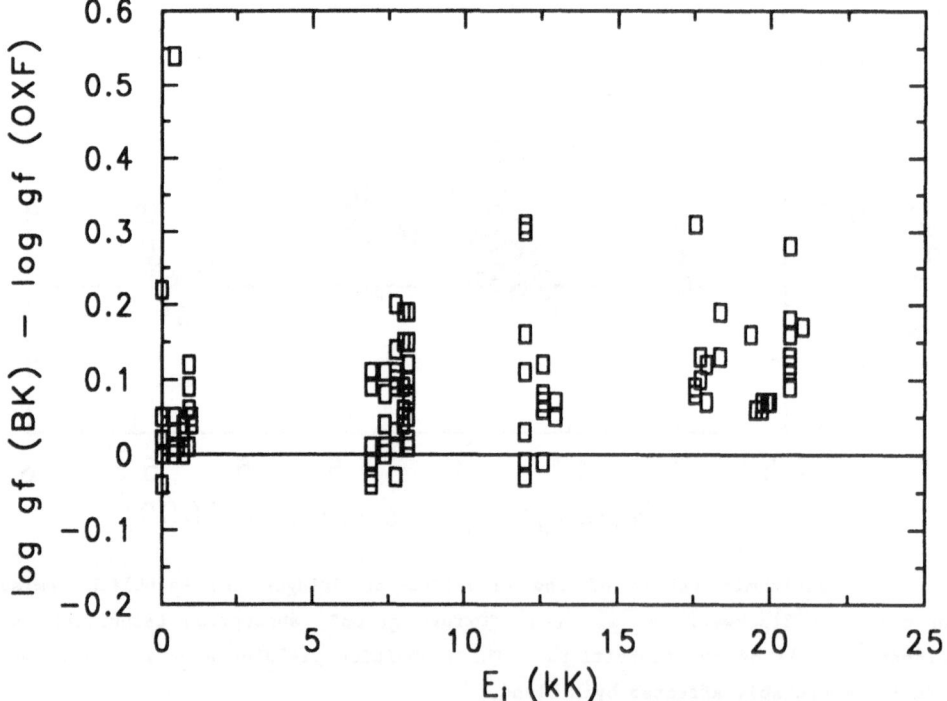

Fig. 4. Same ratios as in Fig. 2, but plotted versus lower energy level of the transitions.

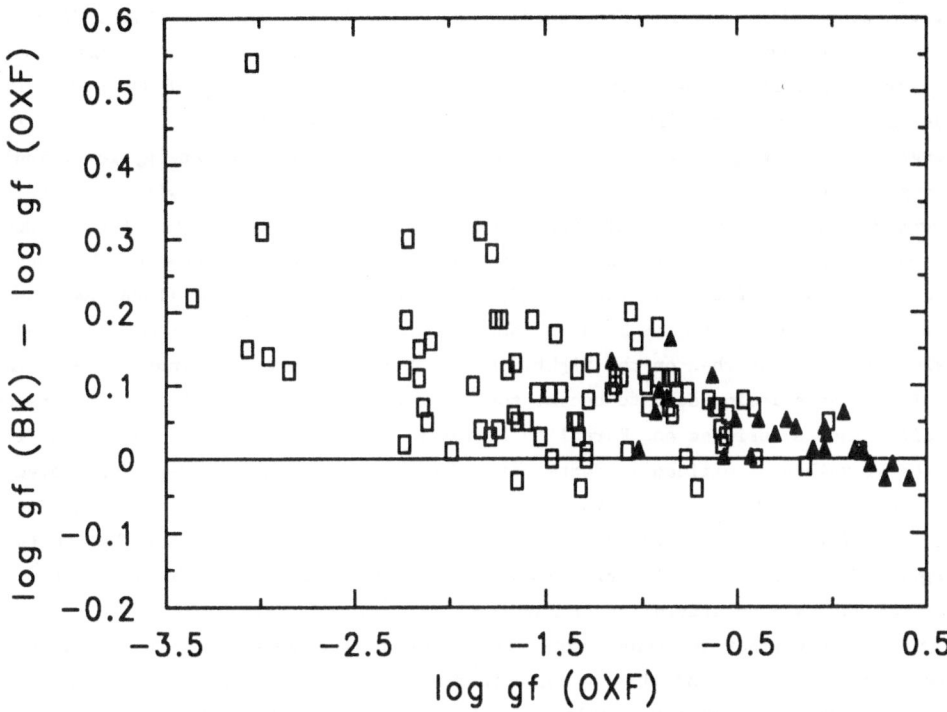

Fig. 5. Same ratios as in Fig. 2, but plotted versus the transition strength, log gf (Oxf.). The triangles contain the high accuracy data of Bridges and Kornblith.[13]

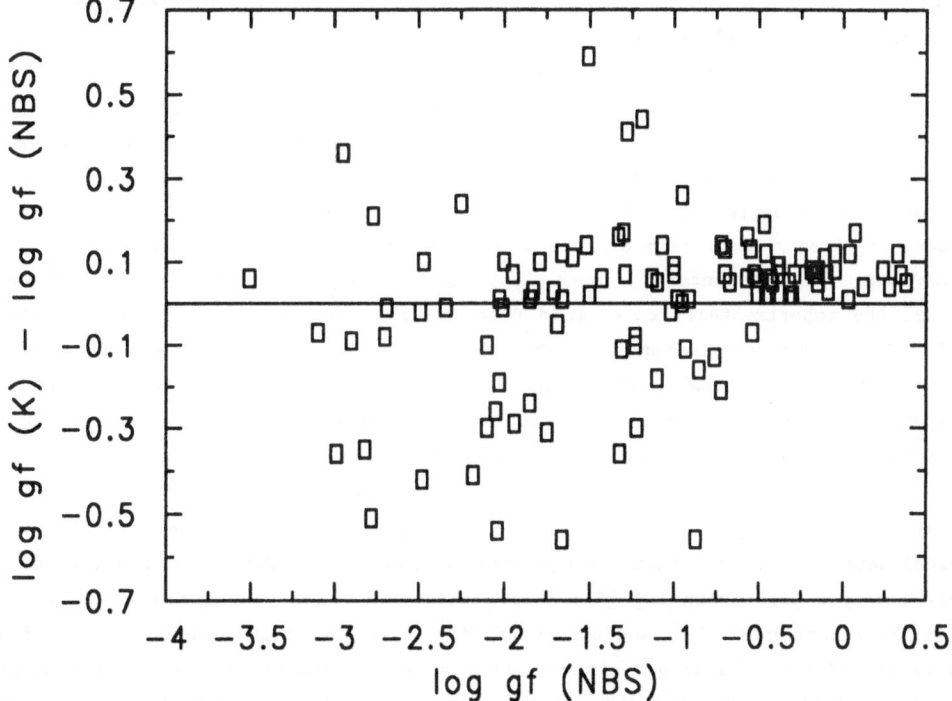

Fig. 6. Logarithmic ratios of gf-values calculated by Kurucz and the experimental data selected for the NBS compilation versus the transition strength log gf (NBS).

transition probability and is often used in the literature. In a series of figures, we have plotted the log gf ratios versus wavelength (Fig. 2), versus upper energy level (Fig. 3), versus lower energy level (Fig. 4) and versus log gf (Fig. 5). The various plots should readily reveal any possible systematic trends with either wavelength, upper energy level, etc. A systematic trend with upper energy level would, for example, indicate a problem with the emission data, while a trend with lower energy level would point to possible deficiencies in the absorption data. No clear evidence of any such trends exists. But there is definitely a gradual increase in the scatter between the two sets of data as the log gf, i.e., the strength of the lines, decreases. This is clearly seen in figure 5, which shows a much wider scatter in the data on the left side of the graph than on the right. Also, there is a gradual tendency for either the small-log gf emission data to become too large or for the absorption data to become too small. Indeed, Bridges and Kornblith estimate that the accuracy of their data with small log gf values significantly decreases. Their highest accuracy data (estimated to be accurate within ±10%) are contained in the ratios given as solid triangles in Fig. 5 and are all for lines of large oscillator strengths. It is seen that 80% of these data agree within ±15% (a Δ log gf of 0.1 corresponds to 25%) with those of Blackwell et al., who estimate uncertainties of ±5% for their absolute data.

A general assessment of these experiments shows, furthermore, that all critical factors of the emission and absorption methods--such as the existence of partial LTE for measurements on a relative scale--have been considered and taken care of by the authors. At the level of refinement achieved in these experiments, we consider these methods to be among the most reliable. Also, the authors' error estimates appear to be quite complete as well as realistic.

On the whole, the spectrum of neutral iron must be considered as being in relatively good shape. The new NSRDS-NIST tabulation[5] contains about 1950 critically evaluated lines for Fe I, of which about half are estimated to have transition probabilities with accuracies of ±25% or better. Twenty percent of the lines are even estimated to be accurate within 10%.

Proceeding from the spectrum of Fe I to that of Fe II, which is astrophysically just as important, one encounters a much less satisfactory data situation. For this spectrum, the experimentally determined data are less plentiful and of lower quality than for Fe I. The measurements[14-16] are supplemented by--and may be compared to--the comprehensive semi-empirical calculations by Kurucz[17] based on Thomas-Fermi-Dirac type wavefunctions. These semi-empirical calculations should be well suited for the Fe II case, since rather complete experimental energy level data are available.[18]

Fig. 6 shows a comparison between the experimental data selected for the most recent critical compilation by the NIST (NBS) data center (labelled NBS) and Kurucz's semi-empirical data. It is seen that the agreement is generally within ±50% (Δ log gf≈0.18) for the stronger lines with log gf's larger than -0.8. For weaker lines the scatter between the experimental data and theory rapidly increases, rising up to factors of 4. It is estimated that the larger part of the scatter has its origin in the theoretical data, since similar comparisons among the experiments show appreciably less scatter (comparable to the situation of Figs. 2-5). Therefore, Kurucz's theoretical data were only sparingly utilized in the NBS compilation--where a main goal was to compile only data with accuracies estimated to be better than ±50%.[19]   The new NSRDS-NIST (NBS) tables on Fe II contain about 650 lines, of which only 18% are estimated to be accurate within ±25%, while the rest carries accuracy estimates of ±50%, which is a significantly smaller proportion than for the earlier discussed case of Fe I.

The accuracy situation is vastly better for such key transitions as the resonance lines of the alkalis. For example, for the Li I and Na I resonance lines Gaupp et. al.[20] achieved accuracies of 0.2% with a very carefully designed and executed lifetime measurement.   Recent calculated results,[21,22] obtained with sophisticated theoretical approaches, differ by only 0.5%, as Table I shows. However, it may be significant that these as well as numerous earlier calculations (all tabulated in Ref. 20) have yielded results that are almost all slightly larger than the lifetime data. The reasons for this systematic difference are still a puzzle;   one possibility could be the non-consideration of small disalignment effects in the lifetime experiment.[23]

Table I.   Comparison of some recent accurate oscillator strength data for the Li I 2s-2p resonance line: Lifetime experiment by Gaupp et al. and calculations by Fulton, Saraph and Peach

Method	Principal Author	Year	Oscillator-strength
Beam-Laser lifetime[20]	Gaupp	1983	0.7416±0.0012
Dirac-Fock approx. approximation[21]	Fulton	1986	0.760
Close coupling calculation[22]	Saraph	1987	0.748
Semiempirical calculation[22]	Peach	1987	0.744

F. The Accuracy of Atomic Transition Probabilities--The Overall Picture.

Turning from these specific examples to the overall situation on the accuracy of transition probability data, we may draw some general conclusions from the extensive material that has been in recent years compiled and evaluated in the NIST (NBS) Data Center on Atomic Transition Probabilities.   Based on this compilation work, we have assembled estimates of typical accuracies in Table II.   For this purpose, we have divided all spectral lines into six groups, i.e., into stronger and weaker lines in each of 3 types of atomic structures.   Clearly, the data for weaker lines--having oscillator strengths $<10^{-2}$--are of much lower quality than for stronger lines.

Table II. Typical Accuracies of Atomic Transition Probabilities

1. <u>Simple atomic structures</u> (1 or 2 electrons outside closed shells, e.g., alkalis, alkaline earths):

Stronger lines ($f > 10^{-2}$)	< 10%	(Th/E)*
Weaker lines	25%	(Th/E)

2. <u>Moderately complex atoms</u> (lighter elements with several electrons outside closed shells, e.g., C, N, O, Al)

Stronger lines	10-15%	(Th/E)
Weaker lines	50%	(E/Th)

3. <u>Complex atomic structures</u> (e.g., Fe, Co, Ni):

Stronger lines	< 25%	(E/Th)
Weaker lines	50% to factors of 2 or more	(E/Th)

*Numerical data are primarily from Theory (T), lesser amounts from Experiment (E)

In Table III, typical data accuracies are shown for the various experimental techniques. The theoretical approaches are not included here, since it is difficult to provide typical accuracy ratings. For example, multi-configuration self-consistent field calculations have produced excellent results for stronger lines of simple atomic structure, and often agree with the best experimental data to within a few percent. But occasionally one still finds unexplained disagreements with high-quality experiments that are of the order of ±25%. The same theoretical approach applied to heavier, more complex atoms sometimes produces appreciable differences with experimental data for weaker as well as stronger lines, but produces good agreement in many other cases. (It should be noted that in almost all original theory papers no accuracy estimates are given.)

Table III. Typical Accuracies Achieved by Principal Experimental Methods

<u>Method</u>	<u>Estimated Accuracy</u>	
	"Best"	"Typical"
Emission	3%(relative)	10-50%
Absorption	1%(relative)	10-30%
Anomalous Dispersion	3%(relative)	10-30%
Lifetime	0.2%(absolute)	5-20%

In conclusion, we should mention that the NIST (NBS) Data Center on Atomic Transition Probabilities has just published two additional volumes of evaluated data[4,5] for approximately 18000 allowed and forbidden lines of the eight elements of the Fe-group--Sc, Ti, V, Cr, Mn and Fe, Co, Ni. The tables include all stages of ionization for which reliable, i.e., better than ±50%, data are available. These data volumes supersede earlier, smaller scale NIST (NBS) compilations on these elements[24-27] and supplement two data volumes covering the first twenty elements.[2,3]

References

1. J. R. Fuhr, B. J. Miller, and G. A. Martin, "Bibliography on Atomic Transition Probabilities," Nat. Bur. Stds. Spec. Pub. 505, U.S. Government Printing Office, Washington, D.C. (1978); Supplement 1 (1980); plus a master list of literature references up to the present time (late 1988), kept at the NIST (formerly NBS) Data Center on Atomic Transition Probabilities.

2. W. L. Wiese, M. W. Smith, and B. M. Glennon, "Atomic Transition Probabilities-Hydrogen through Neon," Vol. I, Nat. Stand. Ref. Data Sec., Nat. Bur. Stand. (U.S.) $\underline{4}$ (1966).

3. W. L. Wiese, M. W. Smith, and B. M. Miles, "Atomic Transition Probabilities-Sodium through Calcium," Vol. II, Nat. Stand. Ref. Data Ser., Nat. Bur. Stand. (U.S.) $\underline{22}$ (1966).

4. G. A. Martin, J. R. Fuhr, and W. L. Wiese, "Atomic Transition Probabilities--Scandium through Manganese," J. Phys. Chem. Ref. Data $\underline{17}$, Suppl. 3 (1988).

5. J. R. Fuhr, G. A. Martin, and W. L. Wiese, "Atomic Transition Probabilities--Iron through Nickel," J. Phys. Chem. Ref. Data $\underline{17}$, Suppl. 4 (1988).

6. See, e.g., K.-N. Huang, At. Data Nucl. Data Tables $\underline{34}$, 1 (1986) and references quoted therein.

7. For a detailed description of the methods, see, e.g., W. L. Wiese, Chapter 25 in "Progress in Atomic Spectroscopy," W. Hanle and H. Kleinpoppen, Eds., Plenum Press, New York (1979).

8. D. E. Blackwell, P. A. Ibbetson, A. D. Petford, and M. J. Shallis, Mon. Not. R. Astron. Soc. $\underline{186}$, 633 (1979).

9. D. E. Blackwell A. D. Petford, and M. J. Shallis, Mon. Not. R. Astron. Soc. 186, 657 (1979).

10. D. E. Blackwell, A. D. Petford, M. J. Shallis, and G. J. Simmons, Mon. Not. R. Astron. Soc. $\underline{191}$, 445 (1980).

11. D. E. Blackwell, A. D. Petford, M. J. Shallis, and G. J. Simmons, Mon. Not. R. Astron. Soc. $\underline{199}$, 43 (1982).

12. D. E. Blackwell, A. D. Petford, and G. J. Simmons, Mon. Not. R. Astron. Soc. $\underline{201}$, 595 (1982).

13. J. M. Bridges and R. L. Kornblith, Astrophys. J. $\underline{192}$, 793 (1974).

14. W. Whaling (to be published).

15. J. M. Bridges, "Contrib. Papers-11th Internat. Conf. Phen. Ioniz. Gases," 418 (I. Stoll, Ed., Czech. Acad. Sci., Inst. Phys., Prague, Czech., 1973).

16. S. Kroll and M. Kock, Astron. Astrophys, Suppl. Sec. $\underline{67}$, 225 (1987).

17. R. L. Kurucz, Smithson. Astrophys. Observ. Spec. Rept. 390 (1981).

18. S. Johansson, Phys. Ser. $\underline{18}$, 217 (1978).

19. The criterion used by the NBS data center for utilizing Kurucz's calculations was to include only strong (log gf >- 1.0) observed lines. These must have also configurationally relatively pure upper and lower energy levels (purity of level≥75%, as taken from the principal eigenvector components of Kurucz).

20.   A. Gaupp, P. Kuske, and H. J. Andrä, Phys. Rev. A $\underline{26}$, 3351 (1982).

21.   T. Fulton and W. R. Johnson, Phys. Rev. A $\underline{34}$, 1686 (1986).

22.   H. E. Saraph and G. Peach, Informat. Quarterly f.Atomic Process. a. Appl. $\underline{29}$, 2 (1987), Daresbury Lab., U. K.

23.   T. Fujimoto, C. Goto, Y. Uetani, and K. Fukuda, Jap. J. Appl., Phys. $\underline{24}$, 875 (1985).

24.   M. W. Smith and W. L. Wiese, J. Phys. Chem. Ref. Data $\underline{2}$, 85 (1973).

25.   W. L. Wiese and J. R. Fuhr, J. Phys. Chem. Ref. Data $\underline{4}$, 263 (1975).

26.   S. M. Younger, J. R. Fuhr, G. A. Martin, and W. L. Wiese, J. Phys. Chem. Ref. Data $\underline{7}$, 495 (1978).

27.   J. R. Fuhr, G. A. Martin, W. L. Wiese and S. M. Younger, J. Phys. Chem. Ref. Data $\underline{10}$, 305 (1981).

Accuracy of Excitation and Ionisation Cross Sections

K. Butler

Institut für Astronomie und Astrophysik,
Scheinerstr. 1,
D–8000 München 80.

Abstract

The availability and accuracy of atomic collisional excitation, collisional ionisation and photoionisation cross sections is discussed. Particular attention is called to the work of Bell et al. (1983), Gallagher and Pradhan (1985) and Seaton and co-workers (1987) who provide references or data for a large number of species.

1 Introduction

Of the two available methods for the determination of atomic cross sections, the experimental technique has the advantage that we are actually dealing with real atoms but the disadvantages are that the number of participating particles is hard to determine and it is difficult to establish what process has actually occurred. Theoreticians on the other hand have to deal with approximate atoms but there are then no normalisation problems (providing the calculations are done correctly) and the various processes (channels) are clearly defined. How these various considerations affect collisional ionisation, collisional excitation and photoionisation cross sections is discussed in the following sections. Particular attention is given to the possible accuracy in each case.

2 Collisional Ionisation

Theory has most difficulty with the calculation of collisional ionisation cross sections due to the presence of three particles in the asymptotic region. Although Seaton and Rudge (1965) wrote down an analytic expression for the wave functions in this region, it cannot be used due to the presence of an unknown charge-like quantity in their formula. Various approximations have been made, but they are all unsatisfactory in that, although they became exact in the limit of high energies, there is no way a priori to say how good the calculations are at lower energies, nor is there any way to improve them systematically.

However, for experimentalists the three particles are an aid since by performing coincidence experiments they can deal with the second above-mentioned problem namely that of deciding what process has been observed. This advantage is somewhat offset through the difficulty of producing sufficient atoms/ions to measure. Only of order 100 measurements have been performed, exclusively of ground state cross sections. Some recent measurements of total cross sections from the work of Crandall et al. (1986) for B^{2+} and O^{5+} and Belic et al. (1987) for Hg^+ are illustrated in figs. 1 to 3. They are compared with various theoretical estimates. The reader is referred to the appropriate papers for details. The experimental error bars indicate an accuracy of better than 10%. While there is good agreement for the lighter species, the Hg^+ cross section obviously still presents problems. This is, of course, generally true and not only for collisional ionisation.

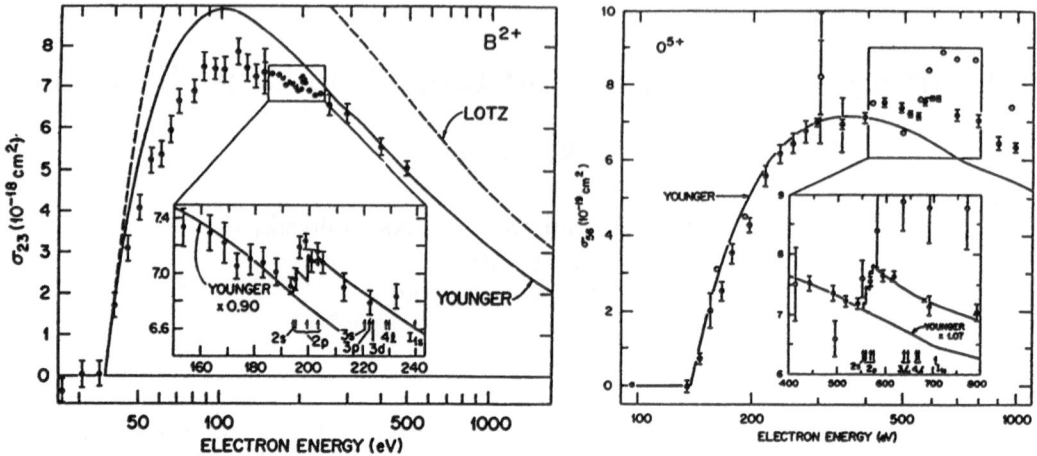

Figure 1: Collisional ionisation cross section for B^{2+} from Crandall et al. (1986) compared with theory of Lotz (1968) and Younger (1980)

Figure 2: Same as fig. 1 but for O^{5+}.

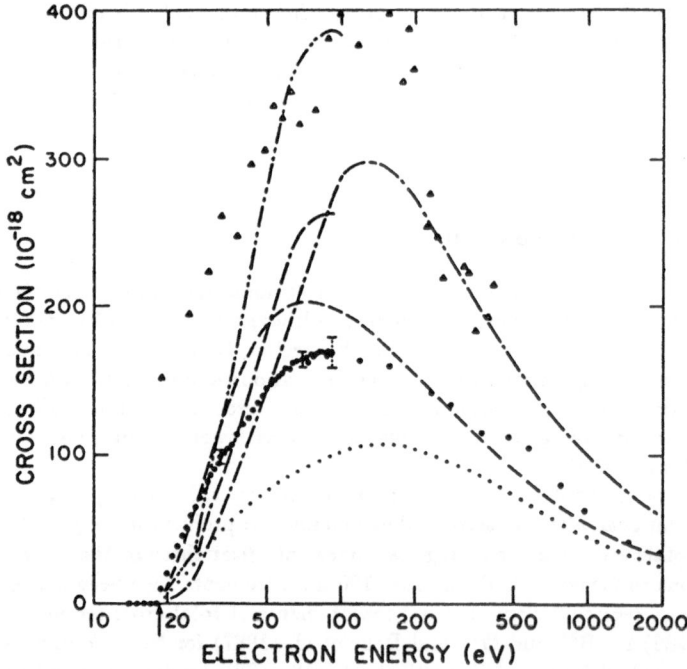

Figure 3: Collisional ionisation cross section for Hg^+ from Belic et al. (1987) (solid circles). See paper for details of other symbols

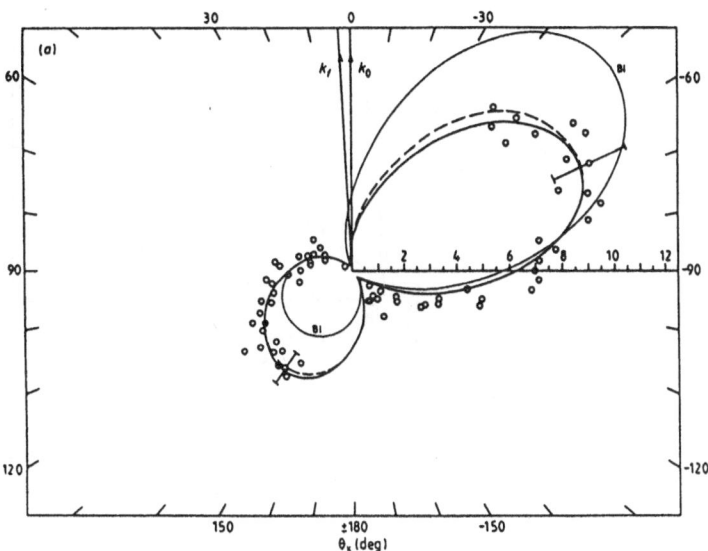

Figure 4: Polar plot in (au) of the TDCS of H as a function of ejected electron angle θ_k from Curran and Walters (1987). Circles: experiment from Ehrhardt et al. (1985). Squares: Lohmann et al. (1984). Results for various ejected electron energies E_k, scattering angles θ_f and momentum transfers, q: (a) $E_k = 5$ eV, $\theta_f = 3°$, q = 0.27 au

Because they provide much additional information, the differential cross sections are a very sensitive test for theoretical treatments. This is seen clearly in the work of Curran and Walters (1987) for hydrogen and Mota Furtado and O'Mahoney (1988) for helium, shown in figs. 4 and 5 respectively. The agreement between the experimental points and the theoretical curves is obvious. Shown are the triple differential cross sections (TDCS) plotted for given incoming electron energy, at various ejected electron energies and scattering angles. The largest differences occur where the experimental results are least reliable. Earlier attempts completely failed to reproduce the back-scattering region (positive θ in the plots) in particular but the newest results can reproduce the experimental values to within the error limits. The main point being that these methods and others (e.g. Jakubowicz and Moores (1981) or Bartschat and Burke (1987)) can be generalised to other, more complex systems.

While we must wait for such calculations, S.M. Younger (see also figs. 1 and 2) has computed numerous ground state cross sections for a wide variety of ions. A recent reference is to be found in the list. Bell et al. (1983) have critically assessed the data available for the lighter ions and recommend cross sections, fit them and tabulate convenient rate coefficients. Some examples may be seen in fig. 6.

There are very few calculated and no experimental values for excited states. A convenient method is provided by the Z-scaled formulae of Clark and Sampson (1984), although the accuracy is uncertain and resonant contributions are lacking. Comparison with ground state calculations by Jakubowicz and Moores (1981) indicated that the Z-scaled data are reliable for relatively low Z values. Another popular means of estimating these excited state (and ground state) cross sections has been the semi-empirical formula of Lotz (1968). However, as is clear from figs. 1 and 2. this generally provides an overestimate.

3   Collisional Excitation

Absolute experiments are difficult for such processes especially for ions. Gallagher and Pradhan (1985) quote only measurements for

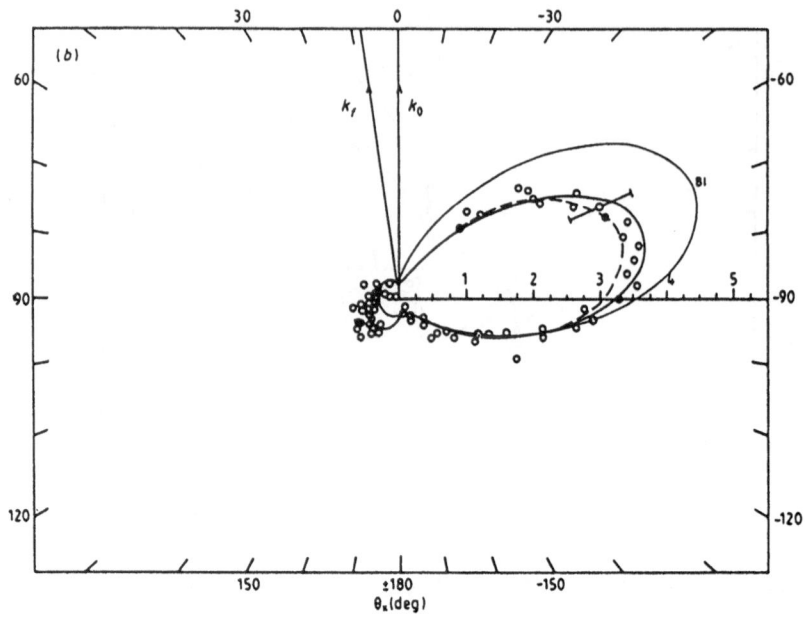

(b) $E_k = 5$ eV, $\theta_f = 8°$, q = 0.61 au

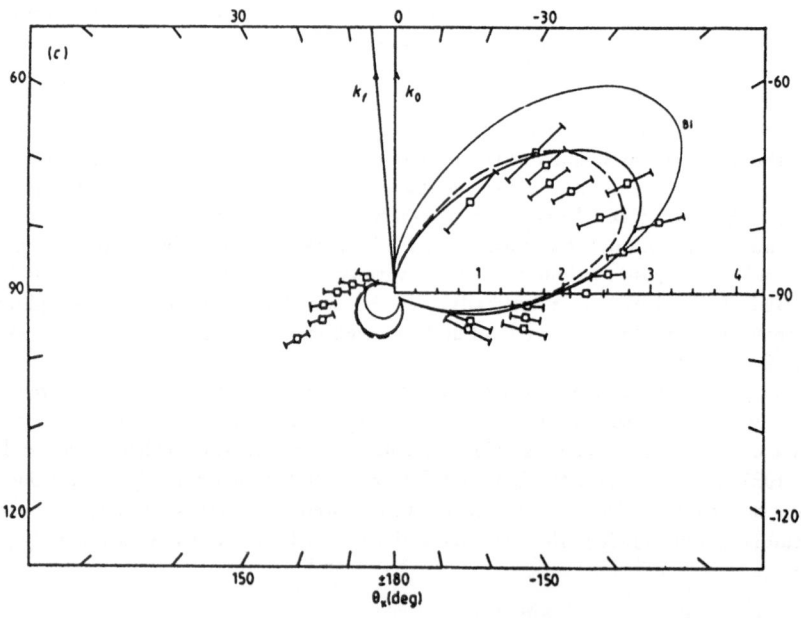

(c) $E_k = 10$ eV, $\theta_f = 5°$, q = 0.42 au

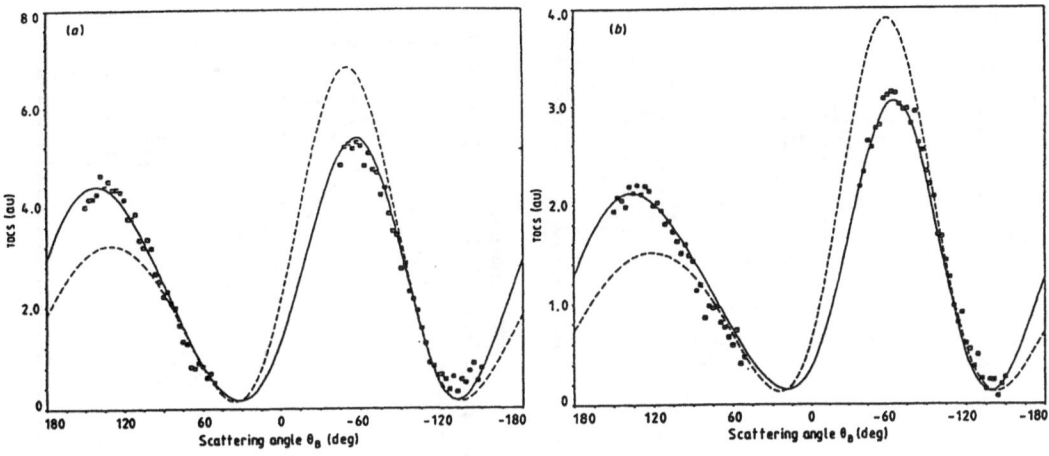

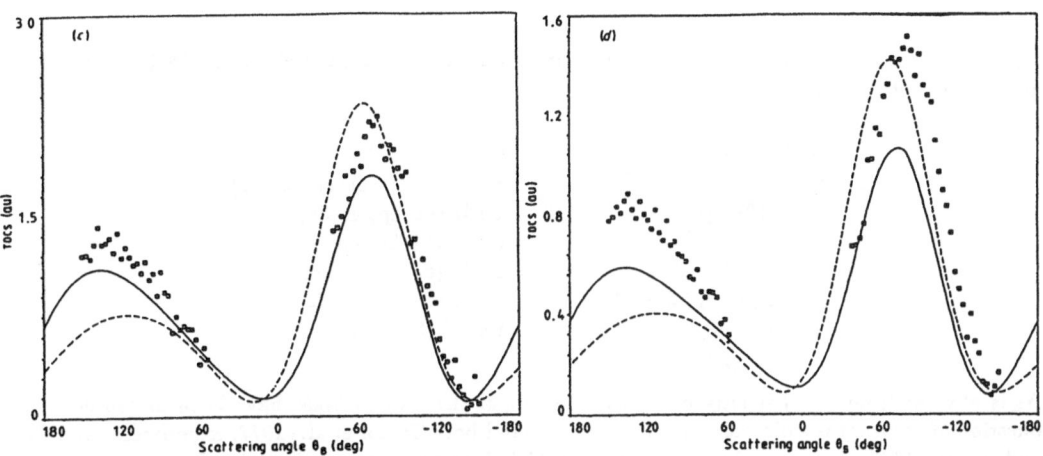

Figure 5: TDCS of He as a function of ejected electron angle θ_B from Mota Furtado and O'Mahony (1988). The squares are experiment of Müller-Fiedler et al. (1985). Plots are for scattering angle θ_A (a) = 4°; (b) = 6°; (c) = 8°; (d) = 10°

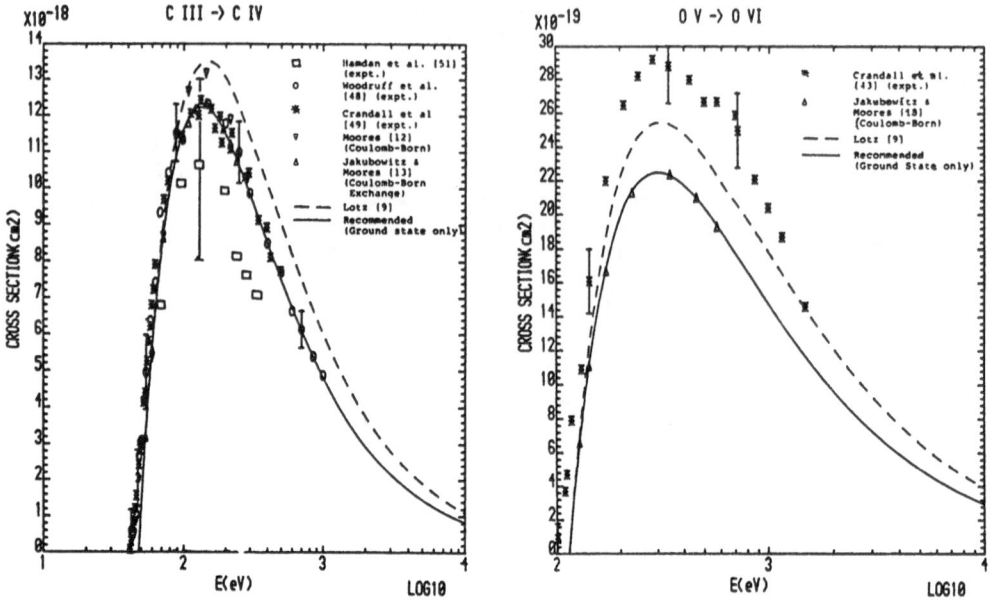

Figure 6: Collisional ionisation cross sections of C III and O V from Bell et al. (1983). The paper gives details of the references shown

He II (1s–2s, 1s–2p)	Ca II (4s–4p)
Li II (1^1S–2^3P)	Sr II (5s–5p, 5s–6s, 5s–5d)
Be II (2s–2p)	Zn II (4s–4p, 4s–5s)
C IV (2s–2p)	Cd II (5s–5p, 4d–5s)
N V (2s–2p)	Ba II (6s–6p, 6s–6d, 6s–7s)
Mg II (3s–3p, 3s–4s)	Hg II (6s–6p, 6s–7s)
Al III (3s–3p)	Hg III (5d–6s).

As is obvious from this list only cross sections out of the ground state have been measured. The question of accuracy is a difficult one. On the one hand both Taylor et al. (1977, experiment) and Gau and Henry (1977, theory) claim an accuracy of 10% for their cross sections for the 2s–2p transition in C IV and their results do indeed agree to this level of accuracy (fig. 7). However, there is only 17% agreement for the same 2s–2p transition in Be II between the experimental result of Taylor et al. (1980) and the best calculated value, even though the expected errors from both are less than 10%.

It is possible to check the internal consistency of the calculations, for example by comparing energy levels, f-values and other observed parameters or by increasing the complexity of the calculation and looking for convergence in the desired quantities.

Gallagher and Pradhan have assessed the literature available to April 1985 and rate the data as A (better than 10%); B (better than 20%); C (better than 30%); D (better than 50%); E (worse than 50%); and U (uncertain). These authors provide the following histogram of their ratings (fig. 8). As can be seen from the diagram relatively few cross sections are known with an accuracy better than 20%. Unfortunately this situation will not alter dramatically in the near future since a great deal of man and computer power must be expended to achieve such results.

A summary of the most used methods may prove helpful. They are listed in order of increasing accuracy. Gallagher and Pradhan (1985) discuss these in much greater detail than can be done here.

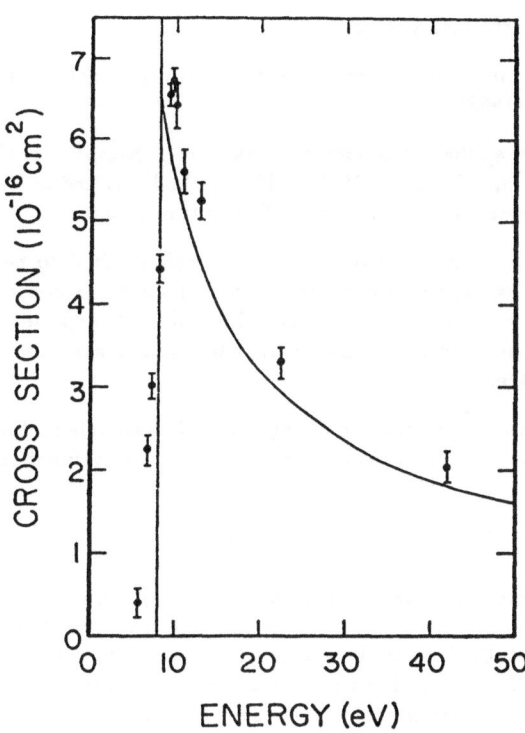

Figure 7: Cross section for 2s–2p excitation of C IV vs electron impact energy. Solid curve: Gau and Henry (1977), points: Taylor et al. (1977)

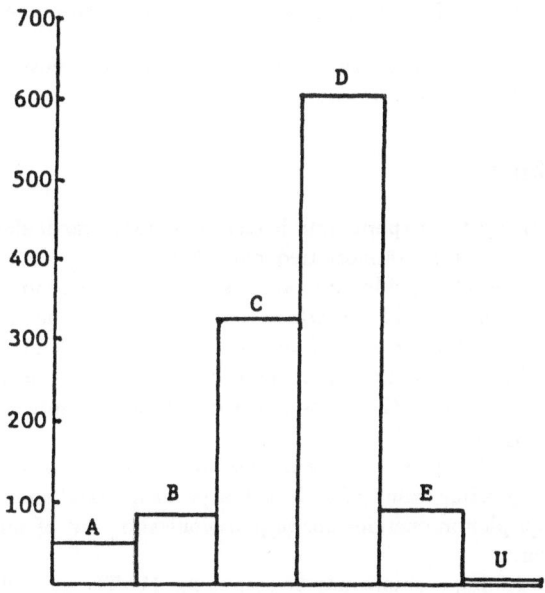

Figure 8: No. of papers with a given accuracy from Gallagher and Pradhan (1985). The symbols are explained in the text

Born applicable only at very high energies

Coulomb-Bethe only for optically allowed transitions. It is probably best known in the form given by van Regemorter (1962).

Coulomb-Born has been applied by numerous workers (e.g. Sampson et al. (a full list of references is given in Gallagher and Pradhan (1985)). The method is suited to highly charged ions and/or high energies and particularly to optically allowed transitions.

Distorted Wave There are many different versions of this method to be found in the literature. Those that are not 'unitarised' are not to be trusted since flux conservation is then no longer guaranteed. In any case the method is only reliable when the species is more than a few times ionised. Most of such calculations ignore the effect that resonances might have on the cross section (see next category).

Close-coupling This is the most accurate of the currently available methods at low energies. The total ion + incoming electron system is represented by a wavefunction Ψ of the form

$$\Psi = \mathcal{A} \sum_{i=1}^{N} \psi_i \theta_i$$

where ψ_i is a wavefunction for the 'target' ion, θ_i a wavefunction for the scattered electron and $\mathcal{A}$ is an antisymmetrising operator to ensure that the total wavefunction obeys the exclusion principle. In the limit that $N \to \infty$ the expansion is exact. The expression may be approximated by taking a finite N. This method takes the coupling between the possible end results (channels) fully into account. This channel-coupling gives rise to resonances in the calculated cross sections (incidentally such resonances appear, in principle at least, in all atomic processes). An example may be seen in fig. 9 which shows the cross section for the ground state fine structure transition in N III by Butler and Storey (1988). The method is limited by the value of N that can be used in the expansion. The largest such calculation published so far is that of Berrington and Kingston (1987) who treated He I while including all terms up to n = 4.

Thus, to summarise, while close-coupling results are preferable and the most accurate very few cross sections are available where this method has been used.

4   Photoionisation

Here again, there are relatively few experiments better than 10%, particularly for ions. Relative cross sections, being easier to obtain, are more frequent. Modern experiments look for coincidences between the photon and photoelectron, older ones used oscillator strength sum rules or the like to put the cross sections on an absolute scale. The results of this procedure can be inaccurate as Butler and Mendoza (1983) showed (fig. 10) for the case of neutral sodium. The theoretical and experimental results agreed at all photoelectron energies when the experimental cross section was reduced by a factor of 0.71. That is to say, although the shape of the experimental cross section was correct, the absolute value was in error by about 40%.

Butler and Mendoza used the close-coupling method in generating their cross section. In this case the restriction on the possible value of N is not a problem since one is rarely interested in photoionisation at very high photon energies nor in photoionisation out of some state into a highly excited state of the next ion.

Another, recent and interesting example may be seen in fig. 11. The ground state photoionisation cross section of He I into various excited states of He II has been compared with corresponding experimental values by Hayes and Scott (1988). Here, the cross section into the n=2 state of He II is shown.

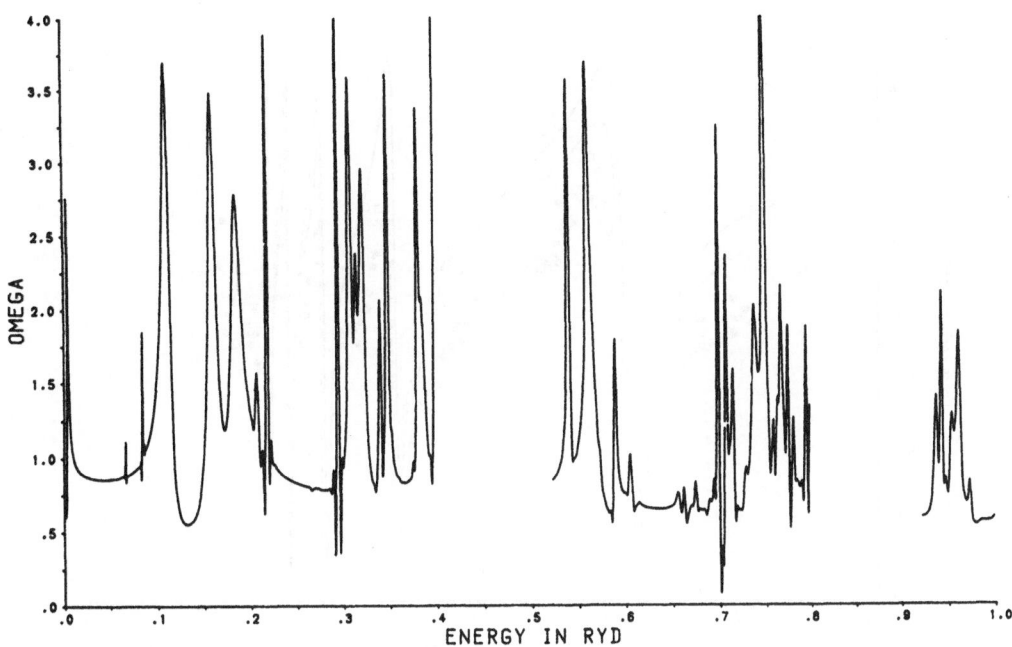

Figure 9: Ground state fine-structure cross section for N III from Butler and Storey (1988). Note the resonances

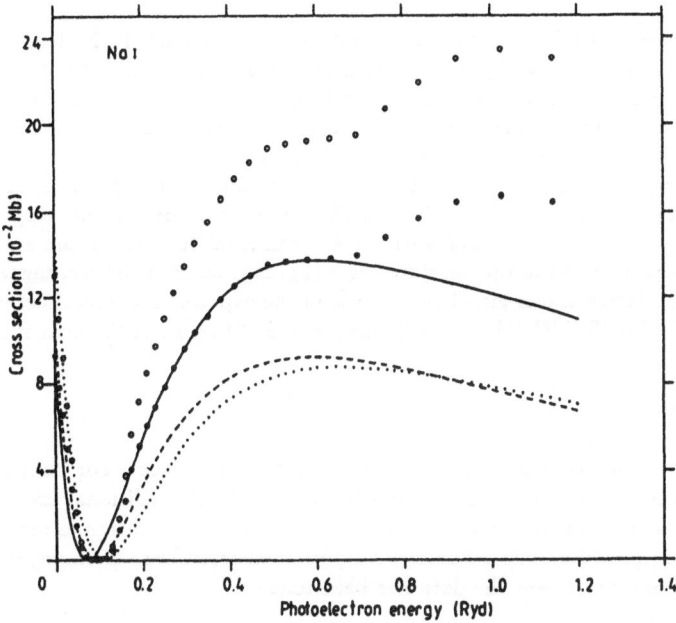

Figure 10: Photoionisation cross section for neutral sodium from Butler and Mendoza (1983) (solid curve) compared with experiment of Hudson and Carter (1967) (open circles) and the same experiment×0.71 (filled circles)

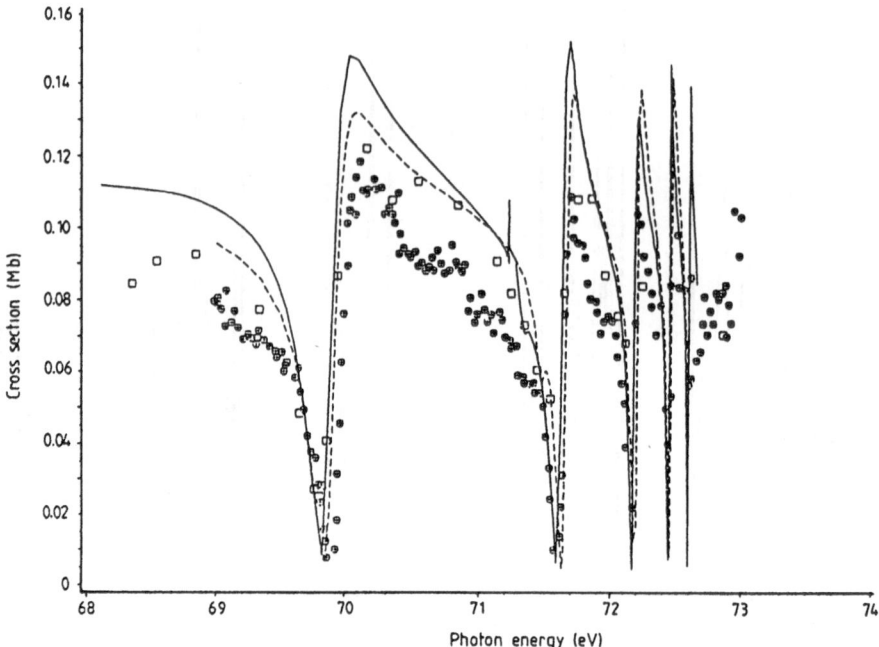

Figure 11: Partial cross section $\sigma_{n=2}$ (Mb) versus photon energy (eV) for He from Hayes and Scott (1988) (solid curve) compared with experiment. Squares: Lindle et al. (1985), crossed circles: Woodruff and Samson (1982)

Recently, Seaton and co-workers (1987) have combined the older RMTRX close-coupling code of Berrington et al. (1978) with a new asymptotic package (Berrington et al. 1987) to provide a means of calculating a large number of atomic parameters. The package requires comparatively little human intervention. The aim was to provide oscillator strengths and photoionisation cross sections for all levels up to n = 10 for all astrophysically important ions of isoelectronic sequences up to that of Ne. In addition, radiative data for relevant ions of Fe were also to be obtained, all data being in LS-coupling. An example is shown in figs. 12 and 13. Taken from Butler and Zeippen (1988), the present close-coupling results are compared with the experimental data of Samson and Angel (1988). Not only is there good agreement in the absolute value (fig. 12), the detailed resonance structure also matches well (fig. 13), though the origin of the 'bump' in the experimental cross section is uncertain.

All data from this 'OPACITY PROJECT' will appear in J. Phys. B. in the near future.

5 Summary

Here, the most interesting references are summarised for convenience. For collisional ionisation Younger (1988), Bell et al. (1983) and Clark and Sampson (1984); for collisional excitation Gallagher and Pradhan (1985); and for photoionisation Seaton et al. (1987). In addition, a recent bibliography by Butler et al. (1988) provides 1500 references for photoionisation data from papers dated 1970 – 1987, although no attempt to assess the data has been made.

Acknowledgements

This work has been supported by the DFG under grant no. Ku 474/13–1. It is also a pleasure to acknowledge the support of the IAU which provided the funding that allowed me to attend the XXth General Assembly in Baltimore.

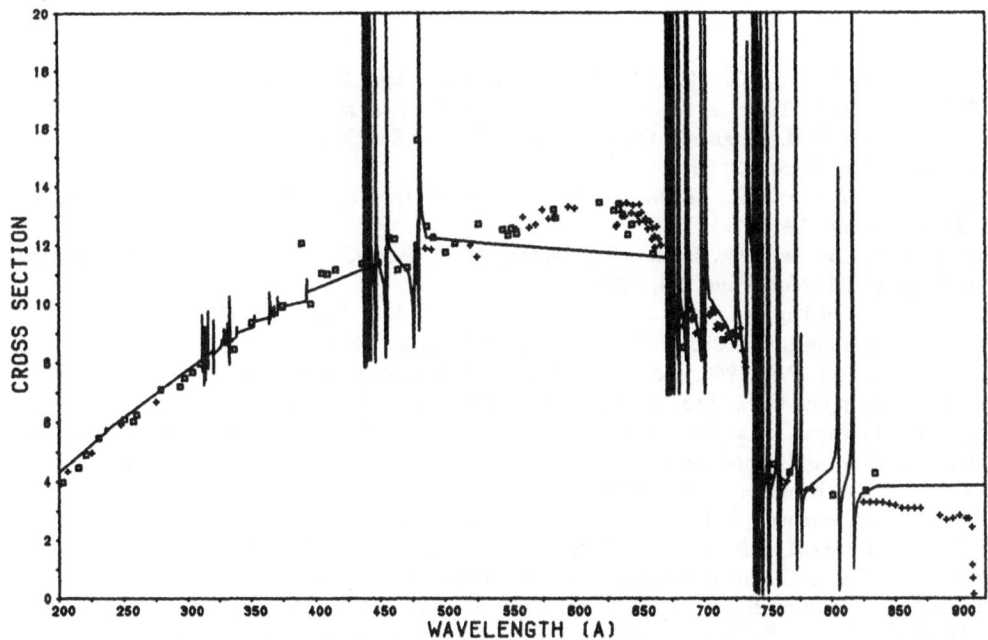

Figure 12: Photoionisation cross section for O I from Butler and Zeippen (1988). Points are experimental data of Samson and Angel (1988)

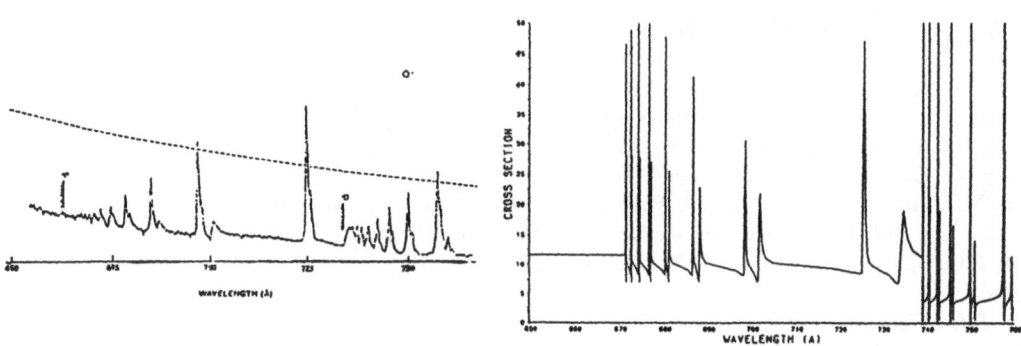

Figure 13: Detail from fig. 12 showing resonance structure. Left; experiment, right; theory

References

Bartschat, K. and Burke, P.G., 1987, *J. Phys. B: At. Mol. Phys.*, **20**, 3191.

Belic, D.S., Falk, R.A., Timmer, C. and Dunn G.H., 1987, *Phys. Rev. A*, **36**, 1073.

Bell, K.L., Gilbody, H.B., Hughes, J.G., Kingston, A.E. and Smith, F.J., 1983,
J. Phys. Chem. Ref. Data, **12**; 891.

Berrington, K.A., Burke, P.G., Le Dourneuf, M., Robb, W.D., Taylor, K.T. and Vo Ky Lan, 1978,
Comp. Phys. Comm., **14**, 367.

Berrington, K.A., Burke, P.G., Butler, K., Seaton, M.J., Storey, P.J., Taylor, K.T. and Yu Yan,
1987, *J. Phys. B: At. Mol. Phys.*, **20**, 6379.

Berrington, K.A. and Kingston, A.E., 1987, *J. Phys. B: At. Mol. Phys.*, **20**, 6631.

Butler, K. and Mendoza, C., 1983, *J. Phys. B: At. Mol. Phys.*, **16**, L707.

Butler, K. and Storey, P.J., 1988, *Mon. Not. R. Astron. Soc.*, in preparation.

Butler, K. and Zeippen, C.J., 1988, *J. Phys. B: At. Mol. Phys.*, in preparation.

Butler, K., Le Dourneuf, M. and Zeippen, C.J., 1988, in *Atomic Data Workshop: Assessment of data
for photo-ionization and photo-excitation and for electron impact excitation of atomic ions*,
ed. Eissner, W. and Kingston, A.E., SERC Daresbury Laboratory.

Clark, R.E.H. and Sampson, D.H., 1984, *J. Phys. B: At. Mol. Phys.*, **17**, 3311.

Crandall, D.H., Phaneuf, R.H., Gregory, D.C., Howald, A.M., Mueller, D.W., Morgan, T.J.,
Dunn, G.H., Griffin, D.C. and Henry, R.J.W., 1986, *Phys. Rev. A*, **34**, 1757.

Curran, E.P. and Walters, H.R.J., 1987, *J. Phys. B: At. Mol. Phys.*, **20**, 337.

Ehrhardt, H., Knoth, G., Schlemmer, P. and Jung, K., 1985, *Phys. Lett.*, **110A**, 92.

Gallagher, J.W. and Pradhan, A.K., 1985, *Jila Data Center Report No. 30*.

Gau, J.N. and Henry, R.J.W., 1977, *Phys. Rev. A*, **16**, 986.

Hayes, M.A. and Scott, M.P., 1988, *J. Phys. B: At. Mol. Phys.*, **21**, 1499.

Hudson, R.D. and Carter, V.L., 1967, *J. Opt. Soc. Am.*, **57**, 651.

Jakubowicz, H. and Moores, D.L., 1981, *J. Phys. B: At. Mol. Phys.*, **14**, 3733.

Lindle, D.W., Ferret, T.A., Becker, U., Kobrin, P.H., Truesdale, C.M., Kerkhoff, H.G. and
Shirley, D.A., 1985, *Phys. Rev. A*, **31**, 714.

Lohmann, B., McCarthy, I.E., Stelbovics, A.T. and Weigold, E., 1984, *Phys. Rev. A*, **30**, 758.

Lotz, W., 1968, *Z. für Physik*, **216**, 241.

Mota Furtado, F. and O'Mahony, P.F., 1988, *J. Phys. B: At. Mol. Phys.*, **21**, 137.

Müller-Fiedler, R., Schlemmer, P., Jung, K. and Ehrhardt, H., 1985, *Z. Phys. A*, **320**, 89.

Rudge, M.R.H. and Seaton, M.J., 1965, *Proc. Roy. Soc. A*, **283**, 262.

Samson, J.A.R. and Angel, G.C., 1988, *SPIE Symposium on X-Ray and VUV Interaction;
Los Angeles, CA, 14 – 15 January 1988*.

Seaton, M.J. and co-workers, 1987, *J. Phys. B: At. Mol. Phys.*, **20**, 6363 – 6476.
N.B. This is a collection of 7 papers on the OPACITY PROJECT.

Taylor, P.O., Gregory, D., Dunn, G.H., Phaneuf, R.H. and Crandall D.H., 1977, *Phys. Rev. Lett.*,
39, 1256.

Taylor, P.O., Phaneuf, R.H. and Dunn, G.H., 1980, *Phys. Rev. A*, **22**, 435.

van Regemorter, H., 1962, *Astrophys. J.*, **136**, 906.

Woodruff, P.R. and Samson, J.A.R., 1982, *Phys. Rev. A*, **25**, 848.

Younger, S.M., 1988, *Phys. Rev. A*, **37**, 4125.

Younger, S.M., 1988, *Phys. Rev. A*, **22**, 111.

ACCURACY OF LINE BROADENING DATA

Milan S. Dimitrijević
Astronomical Observatory
Volgina 7, 11050 Belgrade, Yugoslavia

1. INTRODUCTION

The growing interest in studying line broadening parameters has been, in recent years, motivated by the important role of these data in plasma diagnostics and astrophysics. Many reviews on line broadening in plasmas have been published. Griem (1974, 1975) and Peach (1981) give a detailed report on the theoretical formalism, while experimental data are critically compiled up to 1982 by Konjević and Roberts (1976) and Konjević et al. (1984a) for neutr al non-hydrogenic atoms, and by Konjević and Wiese (1976) and Konjević et al. (1984b) for non-hydrogenic ions. Accuracy of line broadening data for hydrogen is discussed by Mathys (1988). A general report on line broadening induced by non-resonant collisions with neutral atoms is published by Allard and Kielkopf (1982). Information on current work may be found in the proceedings of the Spectral Line Shapes Conferences and in the reports of IAU Commission 14. A bibliography on Atomic Line Shapes and Shifts is published by Fuhr et al. (1972, 1974, 1975, 1978) for the period 1889 through 1978.

2. LINE BROADENING

The absorbing (radiating) system can be described in three different ways: quantum mechanically, semiclassically or, classically. In the pure quantum approach, we usually have a system of non-interacting cells, containing the radiating atom and N perturbers. Now we consider each cell as a giant molecule. To perform a pure quantum mechanical strong coupling calculation on such a cell is very difficult and only a few such calculations exist. For example the strong coupling method was used to determine Stark broadening parameters of the following lines: Li I (2s - 2p) (Dimitrijević et al., 1981), Ca II (4s - 4p) (Barnes and Peach, 1970), Mg II (3s - 3p) (Bely and Griem, 1970; Barnes, 1971), Ca II (3d - 4p) (Barnes, 1971), and Be II (2s - 2p) (Sanchez et al, 1973). Recently, Seaton performed close coupling calculations for 42 transitions in Li-like and Be-like ions C III, O V, Ne VII, Be II, B III, C IV, O VI, Ne VIII (Seaton, 1988) and for the transitions $2s^2\,{}^1S$ - $2s2p^1P^o$, $2s2p^3P^o$ - $2p^2\,{}^3P$, and $2s2p^1P^o$ - $2p^2\,{}^1D$ and ${}^1S$ in C III (Seaton, 1987). Seaton assumes that his results, as solutions of the CC problem which use truncated expansions, are probably correct to within 10 per cent. The modern close coupling methods which have been developed for solving perturber-absorber (emitter) collision problems are described by Burke and Seaton (1971). Extensive packages of computer programs are available and it is clear that these programs could be used to do many more accurate quantum mechanical calculations.

In spite of the existence of the more refined quantum mechanical approach, the semiclassical method is still the most widely used technique for the calculation of line broadening data. In this approach, only the absorbing (radiating) atom is described quantum mechanically. Perturbers are classical particles with a well defined velocity (v) and impact parameter. The system of classical perturbers acts on the quantum mechanical atom via a classical, time dependent interaction potential. The Schroedinger equation which is satisfied by the atomic wave functions is usually solved using second order non- stationary perturbation theory.

Finally, we can use the classical picture. In this the radiating atom is represented by a classical oscillator. The atomic oscillator is perturbed by a system of particles following classical paths. If we assume that the perturbation is adiabatic, and that the important interactions are binary, then a perturber acts on the atomic oscillator via a classical time dependent potential changing only the phase of the oscillator.

3. NEUTRAL ATOM COLLISION BROADENING

While in Stark broadening theory the interparticle interaction is of the Coulomb type, in the case of the neutral-atom absorber collisions, it is not so easy to accurately take into account perturber-atom interactions as a function of their separation. Analyses of line-core shift and width by Hindmarsch et al. (1967, 1970) and Smith (1972), have demonstrated that the van der Waals term alone does not account for the observed effects.

Dimitrijević and Peach (1988) performed an empirical investigation of the validity of the van der Waals formula within Li (2s - 2p), Na (3s - 3p), K (4s - 4p) homologous sequence. In Fig. 1, measured half-half widths w (experiment) are presented for the resonance lines of Li (Gallagher, 1975; Smith and Collins, 1976; Lwin et al. 1977), Na (Mc Cartan and Farr, 1976) and K (Lwin et al. 1977) perturbed by rare gases. The theoretical half-half width is

$$w(theory) \propto C_6^{2/5} \left(\frac{T}{\mu} \right)^{3/10}$$

where C_6 is the van der Waals constant, μ is the reduced mass of the emitter- perturber system, and N and T are the number density of perturbers and the temperature. The coefficient C_6 is obtained by averaging over the degenerate levels, as described by Al-Saqabi and Peach (1987). The polarizability of the rare gas atoms and consequently C_6 increase with increasing atomic mass and it can be seen that the measured widths do in general increase with increasing polarizability. This is an indication that the long range part of the radiator- perturber interaction becomes relatively more dominant.

The results are more dispersed for He and Ne perturbers because the short range part of the interaction is also important and this is not directly related to polarizability at all. It also appears that for heavier perturbers better agreement between theory and experiment is obtained if the van der Waals formula is multiplied by a factor of about 1.25. This can possibly been explained by studying Fig. 2 where accurate potentials for Na-Ne (Peach, unpublished) are plotted as a ratio to their asymptotic limit. This shows that true convergence to the van der Waals limit only occurs for R>50 a:u., a much larger value of R than is often assumed, but that the curves have maxima at much lower values of R.

There have been a number of theoretical efforts to calculate the broadening, particularly of iron by atomic hydrogen, using more refined potentials. For example Brueckner's theory (1971) and tables as a function of effective principal quantum number: the extensive tabulations of Derrider and Rensbergen (1974, 1976) based on the Fermi-Roueff potential, the calculations of Roueff for sodium lines (1974, 1975) the calculation for sodium D lines

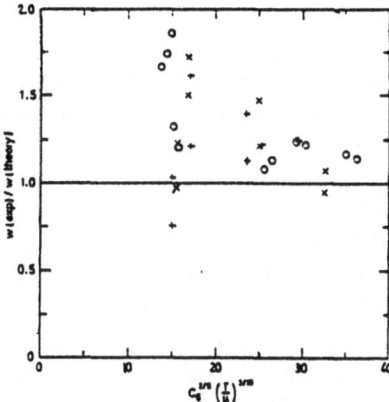

Fig. 1. The ratio w (experiment)/w(theory) as a function of $C_6^{2/5}(T/\mu)^{3/10}$ (C_6 evaluated in atomic units, c.f. Allen (1973)).

by Monteiro et al. (1985). Discrepancies between theory and empirically deduced damping constants of an order of magnitude for solar iron lines are possible. In the case of the sodium D lines, which are very important in studies of the Sun and other stars, the situation is better. All available theoretical and experimental data have been reviewed by O'Mara (1986) and according to the discussion given there, it appears that the calculations by Monteiro et al. (1985) and experimental data by Lemaire et al. (1985) agree well with empirical determinations obtained from the solar spectrum.

Reliable experimental data on the non-resonant neutral atom impact line broadening are compiled by Allard and Kielkopf (1982). Only data of high photometric precision are included and no photographic measurements are used. Also the only measurements given are those for which instrumental effects and other possible sources of broadening have been subtracted.

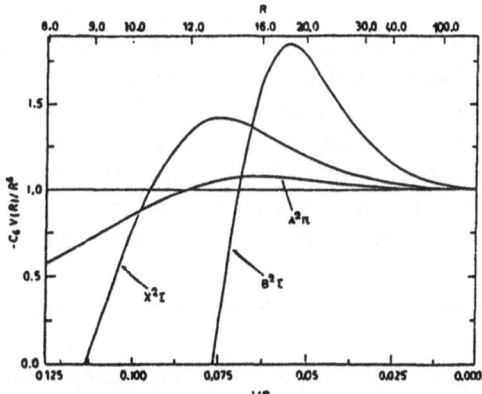

Fig. 2. The adiabatic potentials for the $X^2\Sigma$, $B^2\Sigma$, and $A^2\Pi$ for the system Na-Ne plotted as a ratio to their asymtotic form $-C/R^6$.

4. STARK BROADENING

In the case of hydrogen and hydrogen-like emitter lines the widely used theories are those of Kepple and Griem (1968) and of Cooper, Smith and Vidal (Smith et al., 1969; Vidal et al., 1970, 1971, see also tables in Vidal et al., 1973). In both cases, the quasistatic approximation fails for the ion broadening in the line center. Cooper et al. (1974) have tried to apply the unified theory used for electrons to the ion case. Future improvement can be expected from the Model Microfield Method, developed originally by Brissaud and Frisch (1971). Stehlé et al. (1983) successfully applied this method to calculate hydrogen Balmer line profiles. Recently methods have been developed to account for the effect of ion dynamics (see e.g. Vidal et al., 1973; Griem, 1979, Kelleher, 1988).

Comprehensive calculations of Stark broadening parameters of non-hydrogenic neutral and singly ionized atom lines (helium through calcium and cesium) were published in 1971 and later in 1974 (Benett and Griem, 1971; Jones et al. 1971, Griem, 1974). Using a version of the same code (Griem, 1974) which had been adapted by Dimitrijević for the case of multiply charged ions, data for BrI, GeI, HgI, PbI, RbI, CdI, ZnI (Dimitrijević and Konjević, 1983); OII (Dimitrijević, 1982a); OIII (Dimitrijević, 1980a); CIII (Dimitrijević, 1980b); CIV (Dimitrijević, 1980b, 1988b); NII, NIII, NIV (Dimitrijević and Konjević, 1981a); SIII, SIV, ClIII (Dimitrijević and Konjević, 1982b) and TiII, MnII (Dimitrijević, 1982c) have been computed.

Semiclassical calculations based on the method developed by Sahal-Bréchot (1969a,b) exist for light elements such as C, N, Mg, Si (without the contribution of resonances) (e.g. Sahal-Bréchot and Segré, 1971). Data for alkali-like ions may be found in Fleurier et al. (1977), while Lesage et al. (1983) compare the semiclassical and experimental data for the low-excitation SiII lines. Recently, using the same computer code, extensive calculations for helium (Dimitrijević and Sahal-Bréchot, 1984a,b; 1985a), sodium (Dimitrijević and Sahal-Bréchot, 1985b) and potassium (Dimitrijević and Sahal-Bréchot, 1987) lines have been performed. Data for F I (Vujnović et al. 1983), ArII (Dimitrijević and Truong-Bach, 1986); GaII, GaIII (Dimitrijević and Artru, 1986) also exist. Using this code Lanz et al. (1988) recently computed a complete set of the SiII Stark broadening parameters required for stellar analysis.

Stark width values obtained by the code of Sahal-Bréchot are in general smaller than those obtained using the code according to Griem (1974), due to the symmetrization procedure used by Sahal-Bréchot and to different lower cut- offs. The difference is smaller if the contribution of resonances is taken into account. In the case of MgII and SiII resonance lines, the experimental data (Goldbach et al. 1982; Lesage et al. 1983) agree better with the results obtained using the procedure of Sahal-Bréchot. However, a general conclusion is difficult to obtain (see e.g. Dimitrijević and Sahal-Bréchot, 1985a) since the different assumptions involved in these two versions of the semiclassical method have different validity conditions.

Extensive calculations by Bassalo et al. (1982) obtained by using a different semiclassical method, exist for HeI lines.

Concurrent with the development of Stark broadening theory, numerous experiments were performed in order to check the accuracy of the theoretical calculations, to establish the temperature and density dependence and to provide new data. Konjević and Roberts

(1976), Konjević and Wiese (1976) and, Konjević et al. (1984a,b) performed the critical analysis of all available experiments for neutral and ionized atoms until 1982. In order to select reliable data, authors imposed certain criteria on each experimental paper. The criteria imposed were as follows:

1. An independent and accurate determination of the plasma electron density.
2. A reasonably accurate determination of the electron temperature.
3. A discussion of the other interfering broadening mechanisms and appropriate experimental problems.

One notices the high-accuracy Stark width and shift measurements for HeI, NI, CI and FeI, with uncertainties estimated in the 15-20% range. Konjević et al. (1984a) recommend these data for plasma diagnostic purposes, particularly the HeI lines at 5015.68 Åand 3888.65 Å, for which high-precision measurements with various plasma sources exist and excellent mutual agreement has been found.

However, the scaling of data significantly outside the experimental temperature and electron density ranges may give values of low accuracy. It is often stated that the dependence of line broadening parameters on the temperature is not so critical. However, this dependence often becomes critical just in the astrophysically interesting region where $T \leq 10000K$ (see e.g. Fig. 4).

In general, the agreement between experiments and semiclassical calculations (Griem, 1974) is fairly good for the widths and on average the discrepancy is usually less then $\pm 20\%$ in the case of neutral and singly-ionized emitters. This is well illustrated in Table 1 where average rations of measured to calculated Stark widths and shifts for various elements are given.

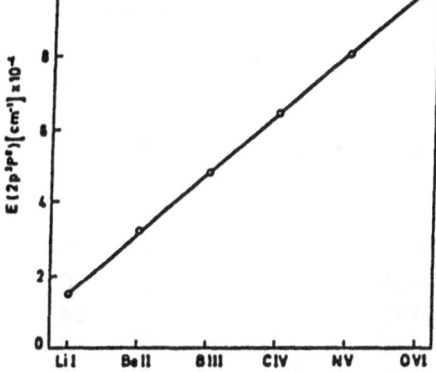

Fig. 3. The energy gap between 2s and 2p levels within Li isoelectronic sequence.

For doubly charged ions the agreemnt is less satisfactory and the semiclassical theory consistently gives larger values than experiment and quantum mechanical calculations (Seaton, 1987, 1988).

If we try to analyse the reasons for such situations, we notice that with the increase of the degree of ionization, the energy gap between perturbing levels increases also (see Fig. 3). As the energy gap increases between perturbing levels, the collisions become more "adiabatic" at the same plasma conditions. The importance of strong collisions increases

Table 1. Average ratios of measured and calculated linewidths (W_m/W_{theor}) for various emitters in the case of various calculations according to Griem (1974) (According to Konjević (1982) and Dimitrijević (1981d). Number of data for W and d are given under n_w and n_w ($n \geq 5$).

ELEMENT	W_M/W_{th}	d_M/d_{th}	n_w	n_d
He I	0.93	1.11	14	14
C I	0.88	1.00	18	9
N I	0.96	0.82	49	26
O I	0.93	1.03	7	5
F I	0.93	1.15	9	8
Ne I	0.84	1.55	17	6
Si I	1.09	0.90	6	8
P I	1.00		10	
S I	1.09		5	
Cl I	1.10		6	
Ar I	0.86	0.81	21	15
Cs I	1.15		19	
C II	1.14		5	
N II	1.09		8	
O II	0.93		13	
Ne II	1.14		12	
Mg II	0.78	1.01	6	6
Al II	1.08		5	
Si II	0.99		6	
S II	1.03		10	
Ar II	1.00	1.15	34	26
Ca II	0.79	0.65	9	7
Ge II	1.05		7	
C III	0.76		7	
O III	0.72		6	
S III	0.73		16	
Cl III	0.74		15	

as well as the importance of resonances and symmetrization. Moreover, the importance of all the other contributions already mentioned increases since perturbers come closer to the emitter due to larger Coulomb attraction making the classical path assumption more questionable.

If we look at a particular spectrum, the semiclassical results are of lower accuracy for the first one or two lines, since in this case the assumptions of the semiclassical approach are not so good due to the significant resonant contribution as well as to the influence of strong and elastic collisions. An example is the MgII resonance line (Fig. 4).

We can see that experimental points fall in two groups separated by more than a factor of two. The data by Goldbach et al. (1982) and Roberts and Barnard (1972) closely follow the results of the quantum mechanical calculations by Barnes (1971) while all other experiments agree closely with calculations of Jones, Benett and Griem (1971). Detailed critical analysis by Konjević et al. (1984) showed that the results of Galbach et al. (1982), and Roberts and Barnard (1972) as well as the quantum mechanical calculations by Barnes (1971) are the most reliable dato for MgII resonance lines.

In Fig. 4 one also notices that the accuracy estimates of experimental data are often too optimistic and that results of several authors do not overlap within their estimated error bars. Consequently, the accuracy of experimental data is not always within error limits estimated by authors.

Another cause of the difference between theoretical and experimental line widths is sometimes the inadequacy of the one-electron model (only one energy level for each nl electron) for line widths of complex spectra (Dimitrijević, 1982a). In such cases, the completeness parameter $\Delta S/S$ (Jones et al., 1971; Griem, 1974) is a good indicator for data less accurate than predicted by the semiclassical theory (Dimitrijević, 1982a). If this parameter differs considerably within a supermultiplet, the accuracy of such data is probably 10-20% lower than expected.

Generally, width data are more reliable than shift data, since shift calculations are more sensitive to the small variations of various parameters. The reason is because shifts are smaller than widths and produced on average, by more distant collisions. This is illustrated by the analysis of Roberts (1968) for the width and shift values convergence as a function of the number of perturbing levels used in the calculation for ArII 4806 Åline. In the case of the shift, even the sign is changed if an insufficient number of perturbing levels is used.

This is also illustrated in Figs. 7 and 8 (Dimitrijević et al., 1981). Here, we show sums of relative contributions to the width and the shift for the various angular momenta l of the colliding electron. We see that in the case of the shift the convergence is not as good as in the case of the width. Consequently, larger computational efforts are needed in order to obtain a good accuracy for the shift.

5. APPROXIMATIVE APPROACHES

Whenever high accuracy line broadening data are not so important, simple, approximate formulae with good average accuracy are very useful. In the case of broadening by collisions with hydrogen, the review article by Allard and Kielkopf (1982) is very useful. In the case of neutral atom Stark broadening there are, for example, simple approaches by Freudenstein and Cooper (1978) and Dimitrijević and Konjević (1986) and for ionized

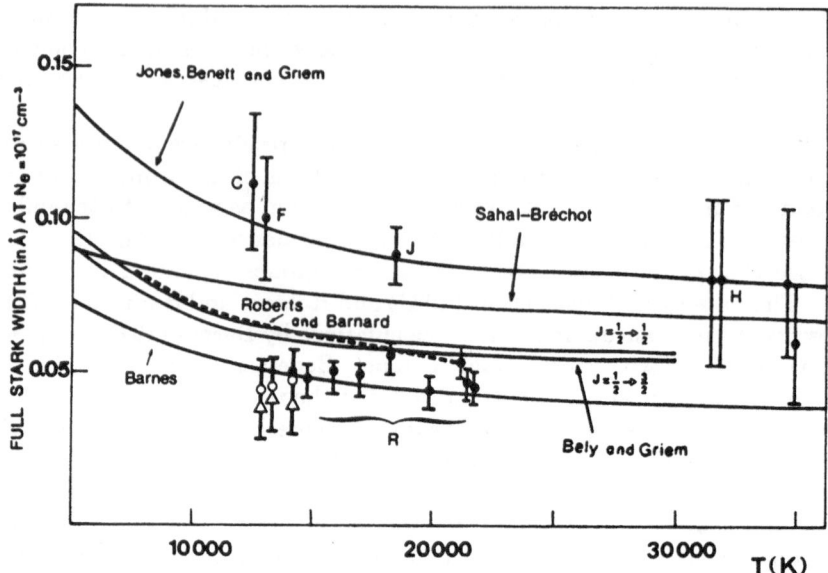

Fig.4. Comparison of Stark width data for the Mg II resonance lines (reproduced from Goldbach *et al.*). Experimental results: C, Chapelle and Sahal-Brechot ; F, Fleurier *et al.* ; J, Jones *et al.* ; H, Hadziomerspahic *et al.* ; R, Roberts and Barnard. The open circles and triangles are the results of the experiment by Goldbach *et al.* for the $J = \frac{1}{2} \rightarrow \frac{3}{2}$ and the $J = \frac{1}{2} \rightarrow \frac{1}{2}$ lines, respectively. The theoretical results, given as the solid and broken lines, are from the quantum mechanical calculations by Barnes, and Bely and Griem , and the semiclassical calculations of Jones *et al.*, Roberts and Barnard, and Sahal-Brechot.

atoms there is the semiempirical method by Griem (1968) and modified the semiempirical formula for widths (Dimitrijević and Konjević, 1980, 1981, 1987) and shifts (Dimitrijević and Kršljanin, 1986). Tables of calculated Stark widths of prominent lines of some doubly and triply-charged ions are given by Dimitrijević and Konjević (1981) and Dimitrijević (1988a).

While for the singly charged ions semiempirical and modified semiempirical formula give similar results, for multiply charged ions the modified semiempirical approach is considerably better. In Table 2 the averaged ratios of experimental and theoretical values for both methods, are given in the case of doubly- and triply-charged ions (Dimitrijević, 1982). We see that the modified semiempirical approach gives on average, for a large number of different emitters, quite reasonable accuracy and may be useful for astrophysical purposes.

Another field where approximate theories can be useful is the realm of heavier elements. Atomic data for heavy element spectra are often insufficient for sophisticated

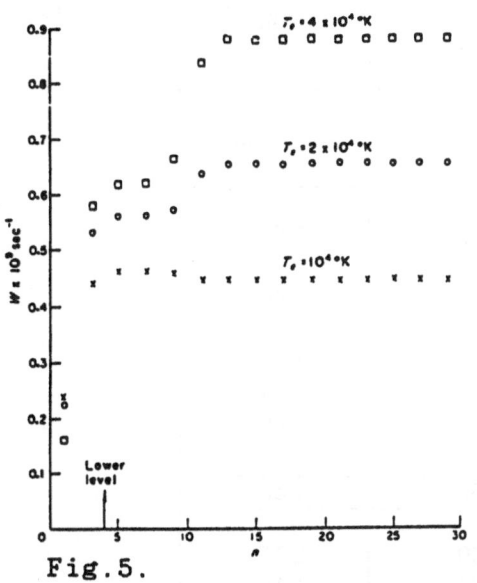

Fig.5.

W vs. n number of perturbing levels used in the calculations (λ4806 Å).

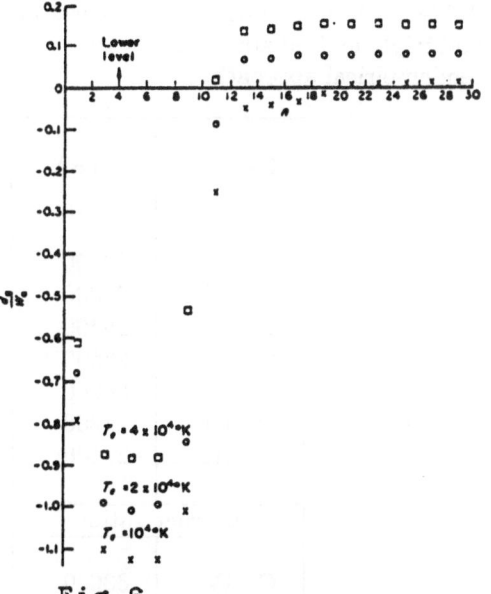

Fig.6.

d, W, vs. number, n, of perturbing levels used in calculation (λ4806 Å).

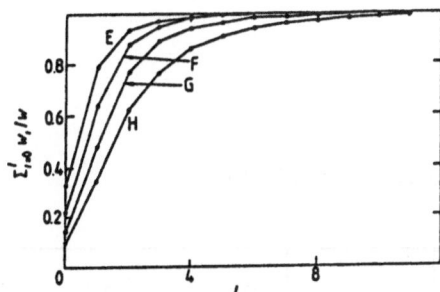

Fig.7. Convergence of the sum $\Sigma_{i=0}^{l} w_i/w$ in the semiclassical approximation as a function of l. The curves E, F, G and H refer to temperatures of 2500, 5000, 10 000 and 20 000 K respectively.

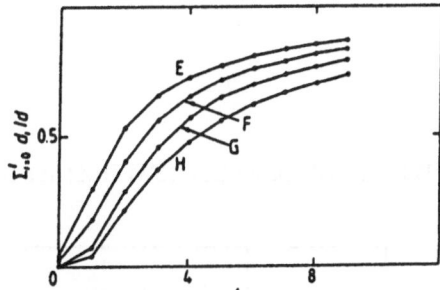

Fig.8. Convergence of the sum $\Sigma_{i=0}^{l} d_i/d$ in the semiclassical approximation as a function of l. Otherwise the same notation.

semiclassical calculations. Moreover, the accuracy of usually applied approximations (e.g.

Table 2. Average ratios of measured (W_m) and calculated linewidths for various doubly and triply ionized atoms. $N_e = 10^{17} cm^{-3}$, W_{SEM}-modified semiempirical approach; W_{SE}-semiempirical approach.

Element	$T_e(K)$	W_m/W_{SEM}	W_m/W_{SE}
C III	60000	1.29	1.21
N III	24300	0.92	1.71
O III	25400	1.05	1.90
Si III	25600	0.67	1.08
S III	28500	1.16	1.65
Cl III	24200	1.01	1.68
A III	21100	0.99	1.57
average ratio:		1.06	1.53
C IV	60000	1.50	2.57
Si IV	25600	0.66	1.15
S IV	28500	0.88	1.65
A IV	20750	0.76	1.24
average ratio:		0.91	1.56

Table 3. Stark widths (FWHM) for heavy elements, $N_e = 10^{17} cm^{-3}$.

Ion	Transition (mult.no.) wavelength Å	$T(K)$	W_{SEM} (Å)	W_{SE} (Å)	W_{SC} (Å)
Ti II	$a^4F - z^4D^o$ (5) 3079.4 Å	2500	0.238	0.230	0.228
		5000	0.168	0.163	0.164
		10000	0.119	0.115	0.125
		20000	0.0840	0.0815	0.0996
Mn II	$a^5D - z^5p^o$ (3) 3464.0 Å	2500	0.264	0.166	0.266
		5000	0.174	0.117	0.190
		10000	0.123	0.0830	0.144
		20000	0.0870	0.0587	0.144

Coulomb approximation) is lower for complex spectra. Additional time consuming sophistications (e.g. Thomas-Fermi model) are needed in order to achieve the accuracy achieved for lighter atoms. In such cases a reliable approximate method can be a useful tool for calculating a quick estimate of the required Stark widths.

As a test for the applicability of the modified semiempirical approach to heavier elements results are presented here for one TiI and one MnII multiplet from Dimitrijević (1982c). The agreement between the simple modified semiempirical approach and the more sophisticated semiclassical calculations is very encouraging and indicates that this simple method can be used to estimate electron widths for heavier elements.

When reliable data do not exist, knowledge about regularities and systematic trends of line broadening parameters can be used for the quick acquisition of new data (especially when high accuracy of the each particular value is not needed) (see e.g. Konjević and Dimitrijević, 1981 and references therein). Similarities and systematic trends can be expected between line broadening parameters of the lines within a given spectrum (multiplet, supermultiplet, transition array), within homologous atoms and ions and within an isoelectronic sequence.

The validity of systematic trends and line broadening data is limited to the plasma conditions for which they are derived and extrapolations are of low accuracy.

REFERENCES:

Allard, N., Kielkopf, J.F.: Rev. Mod. Phys. **54**, 1103

Allen, C.W.: 1973, Astrophysical Quantities, 3rd edn. Athlone Press, London

Al-Saqabi, B.N.I., Peach, G.: 1987, J. Phys. B. **20**, 1175

Barnes, K.S.: 1971, J. Phys. B. **4**, 1377

Barnes, K.S., Peach, G.: 1970, J. Phys. B. **3**, 350

Bassalo, J.M., Cattani, M., Walder, V.S.: 1982, J. Quant. Spectrosc. Radiat. Transfer **28**, 75

Bely, O., Griem, H.R.: 1970, Phys. Rev. A **1**, 97

Benett, S.M., Griem, H.R.: 1971, Calculated Stark Broadening Parameters for Isolated Spectral Lines from the Atoms Helium through Calcium and Cesium, Univ. Maryland, Techn. rep. No. 71-097, College Park, Maryland

Brissaud, A., Frisch, U.: 1971, J. Quant. Spectrosc. Radiat. Transfer **11**, 1767

Brueckner, K.A.: 1971, Astrophys. J. **169**, 621

Burke, P.G., Seaton, M.J.: 1971, Meth. Comp. Phys. **10**,1

Chapelle, J., Sahal-Bréchot, S.: 1970, Astron. Astrophys. **6**, 415

Cooper, J., Smith, E.W., Vidal, C.R.: 1974, J. Phys. B **7**,L101

Derrider, G., van Rensbergen, W.: 1974, Sol. Phys. **34**, 621

Derrider, G., van Rensbergen, W.: 1976, Astron. Astrophys. Suppl. Series **23**, 147

Dimitrijević, M.S.: 1980a, Publ. Astr. Obs. Sarajevo **1**, 215

Dimitrijević, M.S.: 1980b, V ESCAMPING, Dubrovnik, p. 90

Dimitrijević, M.S.: 1982a, Astron. Astrophys. **112**, 251

Dimitrijević, M.S.: 1982b, J.Quant.Spectrosc. Radiat. Transfer **27**, 203

Dimitrijević, M.S.: 1982c, in Sun and Planetary System, eds. W. Fricke, G. Teleki, D. Reidel P.C., Dordrecht, p. 101

Dimitrijević, M.S.: 1982d, in The Physics of Ionized Gases (SBIG-1982, Invited Lectures), ed. G. Pichler, Zagreb 1982, p. 397

Dimitrijević, M.S.: 1988a, Astron. Astrophys. Suppl. Series, accepted

Dimitrijević, M.S.: 1988b, Bull. Obs. Astr. Belgrade **139**, 31

Dimitrijević, M.S., Artru, M.-C.: 1986, XIII Symp. Phys. Ioniz. Gases, Sibenik, p. 317

Dimitrijević, M.S., Feautrier, M., Sahal-Bréchot, S.: 1981, J. Phys. B **14**, 2559

Dimitrijević, M.S., Konjević, N.: 1980, J. Quant. Spectrosc. Radiat. Transfer **24**, 451

Dimitrijević, M.S., Konjević, N.: 1981a, J. Quant. Spectrosc. Radiat. Transfer **25**, 387

Dimitrijević, M.S., Konjević, N.: 1981b, in Spectral Line Shapes, ed. B. Wende, W. de Gruyter, Berlin, New York, 211

Dimitrijević, M.S., Konjević, N.: 1982, J. Quant. Spectrosc. Radiat. Transfer **27**, 203

Dimitrijević, M.S., Konjević, N.: 1983, J. Quant. Spectrosc. Radiat. Transfer **30**, 45

Dimitrijević, M.S., Konjević, N.: 1986, Astron. Astrophys. **163**, 297

Dimitrijević, M.S., Konjević, N.: 1987, Astron. Astrophys. **172**, 345

Dimitrijević, M.S., Krsljanin, V.: 1986, Astron. Astrophys. **165**, 260

Dimitrijević, M.S., Peach, G.: 1988, XIV Symp. Phys. Ioniz. Gases, Sarajevo, p. 321

Dimitrijević, M.S., Sahal-Bréchot, S.: 1984a, J. Quant. Spectrosc. Radiat. Transfer **31**, 301

Dimitrijević, M.S., Sahal-Bréchot, S.: 1984b, Astron. Astrophys. **136**, 289

Dimitrijević, M.S., Sahal-Bréchot, S.: 1985a, Phys. Rev. A **31**, 316

Dimitrijević, M.S., Sahal-Bréchot, S.: 1985b, J. Quant. Spectrosc. Radiat. Transfer **34**, 149

Dimitrijević, M.S., Sahal-Bréchot, S.: 1987, J. Quant. Spectrosc. Radiat. Transfer **38**, 37

Dimitrijević, M.S., Truong-Bach: 1986, Z. Naturforsch. **41a**, 772

Fleurier, C., Sahal-Bréchot, S., Chapelle, J.: 1977, J. Quant. Spectrosc. Radiat. Transfer **17**, 595

Freudenstein, S.A., Cooper, J.: 1978, Astrophys. J. **224**, 1079

Fuhr, J.R., Wiese, W.L., Roszman, L.J.: 1972, Bibliography on Atomic Line Shapes and Shifts (1889 through March 1972), Nat. Bur. Stand. (U.S.) Spec. Publ. **366**, Govt. Print. Office, Washington D.C.

Fuhr, J.R., Roszman, L.J., Wiese, W.L.: 1974, Bibliography on Atomic Line Shapes and Shifts (April 1972 through June 1973), Nat. Bur. Stand. (U.S.) Spec. Publ. **366**, Suppl. 1, Govt. Print. Office, Washington D.C.

Fuhr, J.R., Martin, G.A., Specht, B.: 1975 Bibliography on Atomic Line Shapes and Shifts (July 1973 through May 1975), Nat. Bur. Stand. (U.S.), Spec. Publ. **366**, Suppl. 2, Govt. Print. Office, Washington D.C.

Fuhr, J.R., Miller, B.J., Martin, G.A.: 1978, Bibliography on Atomic Line Shapes and Shifts (June 1975 through 1978), Nat. Bur. Stand. (U.S.), Spec. Publ. **366**, Suppl. 3, Govt. Print. Office, Washington D.C.

Gallagher, A.: 1975, Phys. Rev. A **12**, 133

Goldbach, D., Nollez, G., Plomdeur, P., Zimmermann, J.P.: 1982, Phys. Rev. **25**, 2596

Griem, H.R.: 1968, Phys. Rev. **165**, 258

Griem, H.R.: 1974, Spectral Line Broadening by Plasmas, Academic Press New York

Griem, H.R.: 1975, Adv. Atom. Molec. Phys. **11**, 331

Griem, H.R.: 1979, Phys. Rev. A **20**, 606

Hadziomerspahic, D., Platisa, M., Konjević, N., Popović, M.: 1973, Z. Phys. **262**, 169

Helbig, V., Kelleher, D.E., Wiese, W.L.: 1976, Phys. Rev. A **14**, 1082

Hindmarsch, W.R., DuPlessis, A.N., Farr, J.M.: 1970, J. Phys. B. **3**, L5

Hindmarsch, W.R., Petford, A.D., Smith, G.: 1967, Proc. R. Soc. London, Ser.A **297**, 296

Jones, W.W., Benett, S.M., Griem, H.R.: 1971, Calculated Electron Impact Broadening Parameters for Isolated Spectral Lines from Singly Charged Ions Lithium through Calcium, Univ. Maryland, Techn.Rep. No 71-178, College Park, Maryland

Jones, W.W., Sanchez, A., Grieg, J.R., Griem, H.R.: 1972, Phys. Rev. A **5**, 2318

Kepple, P., Griem, H.R.: 1968, Phys. Rev. **173**, 317

Kelleher, D.: 1988, in Proc. of the Workshop on Spectral Line Formation in Plasmas under Extreme or Unusual Conditions, Nice

Konjević, N.: 1982, in The Physics of Ionized Gases (SPIG-1982, Invited Lectures), ed. G. Pichler, Zagreb 1982, p. 417

Konjević, N., Dimitrijević, M.S.: 1981, in Spectral Line Shapes, ed. B. Wende, W. de Gruyter, Berlin, New York, 1981, p. 211

Konjević, N., Dimitrijević, M.S., Wiese, W.L.: 1984a, J. Phys. Chem. Ref. Data **13**, 610

Konjević, N., Dimitrijević, M.S., Wiese, W.L.: 1984b, J. Phys. Chem. Ref.Data **13**, 649

Konjević, N., Roberts, J.R.: 1976, J. Phys. Chem. Ref. Data **5**, 201

Konjević, N., Wiese, W.L.: 1976, J. Phys. Chem. Ref. Data **5**, 259

Lanz, T., Dimitrijević, M.S., Artru, M.-C.: 1988, Astron. Astrophys. **192**, 249

Lemaire, J.L., Chotin, J.L., Rostas, F.: 1985, J. Phys. B **18**, 95

Lesage, A., Rathore, B.A., Lakicević, I.S., Puric, J.: 1983, Phys. Rev. A **28**, 2264

Lwin, N.D., McCartan, D.G., Lewis, E.L.: 1977, Astrophys. J. **213**, 599

McCartan, D.G., Farr, J.M.: 1976, J. Phys. B **9**, 985

Mathys, G.: 1988, in Elemental Abundances Analyses, eds. S.J. Adelman, T. Lanz, Institut de'l Astronomie de'l Universite de Lausanne, p. 143

Monteiro, T.S., Dickinson, A.S., Lewis, E.L.: 1985, J. Phys. B **18**, 3499

O'Mara, B.J.: 1986, J. Phys. B **19**, L349

Peach, G.: 1981, Adv. Phys. **30**, 367

Roberts, D.E.: 1968, J. Phys. B **1**, 53

Roberts, D.E., Barnard, A.J.: 1972, J. Quant. Spectrosc. Radiat. Transfer **12**, 1205

Roueff, E.: 1974, J. Phys. B **7**, 185

Roueff, E.: 1975, Astron. Astrophys. **38**, 41

Sanchez, A., Blahn, M., Jones, W.W.: 1979, Phys. Rev. A **8**, 774

Sahal-Bréchot, S.: 1969a, Astron. Astrophys. **1**, 91

Sahal-Bréchot, S.: 1969b, Astron. Astrophys. **2**, 322

Sahal-Bréchot, S., Segre, S.: 1971, Highlights of Astronomy **2**, ed. C. de Jager, p. 566

Seaton, M.J.: 1987, J. Phys. B **20**, 4631

Seaton, M.J.: 1988, J. Phys. B **21**, 3033

Smith, G.: 1972, J. Phys. B **5**, 2310

Smith, G., Collins, B.S.: 1976, J. Phys. B, **9**, 2117

Smith, E.W., Cooper, J., Vidal, C.R.: 1969, Phys. Rev. **185**, 140

Stehle, C., Mazure, A., Nollez, G., Feautrier, N.: 1983, Astron. Astrophys. **127**, 263

Vidal, C.R., Cooper, J., Smith, E.W.: 1970, J. Quant. Spectrosc. Radiat. Transfer **10**, 1011

Vidal, C.R., Cooper, J., Smith, E.W.: 1973, Astrophys. J. Suppl. Series **25**, 37

Vujnović, V., Vadla, C., Lokner, V., Dimitrijević, M.S.: 1983, Astron. Astrophys. **123**, 249

ACCURACY OF ABUNDANCES FROM O TO MID B MAIN SEQUENCE STARS

Roberto H. Mendez
Instituto de Astronomia y Fisica del Espacio
C.C. 67, Suc. 28, 1428 Buenos Aires, Argentina

Abstract: These notes start with a description of the non-LTE model atmospheres and line formation programs currently used to determine abundances in hot stars. Some recent results are reported concerning early O and OBN stars, He abundance in h and χ Persei, possible N deficiency in the open cluster NGC 6231, interstellar abundance inhomogeneities, and radial gradients of N and O abundances in our Galaxy. Whenever possible, results obtained by different groups are compared. For the early O stars, the present uncertainties in the He abundances are of the order of 20%, while the abundances for all heavier elements are still very uncertain. The situation for late O and early B stars is better: 20% for He, and about 0.3 dex for C, N, O, Mg, Al, Si. A substantial improvement is expected in the near future, particularly for the earliest O stars.

1. Introduction

I have been asked to report on the accuracy of abundance determinations in main sequence stars having spectral types from early O to mid B. The atmospheres of the hottest main sequence stars are characterized by two main difficulties: departures from local thermodynamic equilibrium (LTE) and the presence of strong stellar winds. The determination of reliable abundances in these atmospheres requires a large computational effort (in some cases at the very limit of, if not beyond, present capabilities). Therefore, we may well ask what are the motivations to make such efforts. We can group the motivations in the following way:

(a) To test theories of stellar structure and evolution. Any well-established chemical anomaly is a challenge to such theories and offers a chance for improvements. The hottest stars near the main sequence are our only way to test (or force) predictions about the initial stages in the evolution of massive stars.

(b) Unevolved stars hot enough to show He lines in their spectra offer a potential way to throw some light on the primordial He problem.

(c) To test theories of Galactic chemical evolution. In particular, young hot stars permit to study abundance gradients in the Galaxy and to check if there exist stellar (and therefore also interstellar) abundance inhomogeneities over relatively small distance scales of the order of 1 Kpc.

Of course, in order to attack all these problems, we need not only good model atmospheres but also good observational material. Fortunately, on the observational side there has been an important progress. For example, the advent of echelle spectrographs equipped with CCD detectors has made it possible (if a telescope of the 4-m class is available) to obtain spectrograms with a spectral resolution of 0.2 Å, covering a spectral range of about 1000 Å, with a signal-to-noise ratio of 50, for stars as faint as $m_v = 13$ or even 14. On such spectrograms it is possible to obtain accurate measurements of equivalent widths as small as 20 mÅ.

In the following sections I will attempt a brief description of the model atmospheres and line formation programs currently in use (and in preparation) to extract abundance information from the optical spectra of hot stars, and will review the application of these tools to the problems listed above. Whenever possible, I will try to compare results obtained by different groups, which is probably the safest way to check what levels of accuracy can be achieved at the present time.

2. Model atmospheres

Most of the current work on abundances in O-type stars is based on the "classical" non-LTE plane-parallel model atmospheres in hydrostatic and radiative equilibrium (Auer and Mihalas,1972; Mihalas, 1972; Kudritzki, 1976; Clegg and Middlemass, 1987). These models give a first approximation to the atmospheric structure (temperature and gas pressure as functions of depth). The modern versions include several slight refinements to the pioneering calculations by Auer and Mihalas; for example, more levels of He and He$^+$ are allowed to depart from LTE.

Combined with suitable line formation programs, to be described in next section, these "classical" non-LTE models already permit to obtain good approximations to the basic atmospheric parameters (log g, T_{eff}, He/H ratio) of all kinds of O-type stars, including Of stars like ζ Puppis (Kudritzki et al., 1983). However, the situation is far from ideal, and there have been several recent attempts to overcome the main limitations of these non-LTE models.

One limitation is the lack in the models of elements heavier than He. Their presence in real atmospheres must produce a certain amount of blanketing due to the blocking of outflowing radiation by bound-bound and bound-free "metal" transitions. The solution would be to produce model atmospheres including many "metal" atomic levels in non-LTE. Although a satisfactory treatment of this problem is not possible yet, the new methods and computer codes developed by Anderson (1985), Werner and Husfeld (1985) and Werner (1986) are important steps in this direction. The exploratory calculations performed up to now (Anderson, 1985; Werner, 1987) suggest that, although there is a large drop in surface temperature, there is not much backwarming in deeper layers. It would seem that the H and He line profiles used to setermine log g, T_{eff} and He/H ratio are not much affected by metal blanketing; by contrast, the effect on the far-UV continuous energy distribution is probably more significant.

Another limitation is the neglect of wind effects. Abbott and Hummer (1985) have introduced an improvement: they take into account the heating of the surface layers produced by the scattering of emerging radiation by the stellar wind back into the photosphere. They refer to this effect as "wind blanketing". Their method is to replace the classical upper boundary condition (no radiation incident on the photosphere) by one in which the incident field is equal to the emergent intensity multiplied by a wavelength-dependent albedo. The calculation of the albedo is described by Abbott and Lucy (1985) and Abbott and Hummer (1985).

More advanced model atmospheres are almost ready. The group of R.P. Kudritzki at the Munich University Observatory is developing a very detailed computer code for radiation driven winds of hot stars (Pauldrach et al., 1986, Kudritzki et al., 1987, Pauldrach, 1987, Puls, 1987). This code, when applied to individual stars, has already shown a reasonable agreement with the observed mass loss rates and terminal velocities, and has provided a convincing explanation of the phenomenon of "superionization" (see in particular Pauldrach, 1987).

Furthermore, the Munich group is developing a so-called "unified model atmosphere", in which there is no artificial separation between a hydrostatic photosphere and a supersonic wind envelope. Instead, their detailed wind code, giving the density structure and velocity field along the whole (sub- and supersonic) atmosphere, is combined with a non-LTE model atmosphere code for spherical geometry, giving the temperature structure. In this way, the whole atmospheric structure is described self-consistently, as a function of three basic stellar parameters: T_{eff}, log g and the stellar radius at which T_{eff} and log g are given (this stellar radius fixes the stellar luminosity and mass), plus the chemical composition. Preliminary results (Kudritzki, 1988, Gabler et al., 1988) show that these models provide a much better description of all the spectral characteristics of O- and Of-type stars, including their emission lines and continuous energy distributions. A basis for improved abundance and determinations (as well as many other applications) appears to be well established.

3. Non-LTE line formation programs

Even assuming hydrostatic and radiative equilibrium, the OB star abundance problem in its most correct formulation is almost hopeless: it would be necessary to solve consistently the coupled equations of radiative transfer and of statistical equilibrium for each energy level of each ion of each element present in the atmosphere.

It is usual practice to treat the problem in the following simplified way: First, the best available model atmosphere is selected. This model gives the atmospheric structure (temperature and gas pressure as functions of depth). Next, two fundamental assumptions are made: that we can keep the atmospheric structure fixed, and that we are allowed to solve a system of radiative transfer and statistical equilibrium equations separately for each ion of each element of interest. This implies to assume that each ion can be treated as a trace element that has no influence on the atmospheric structure (except of course in the particular cases of H and He, where the assumption is that the starting model atmosphere is already so good that the introduction of additional statistical equilibrium equations and additional transitions and other refinements for the calculation of occupation numbers will have no significant effects on the atmospheric structure). Once a satisfactory solution is obtained for the level populations, it is possible to calculate the line profiles and equivalent widths needed for a comparison with the observations.

In the previous section we have seen the current status of model atmosphere calculations. In practice, while we wait for the new generation of non-LTE metal-line blanketed models, there are two alternatives: either non-LTE plane- parallel models without metal blanketing, or LTE models with metal blanketing computed in LTE. The choice depends on the spectral type: for early O stars non-LTE effects are clearly more important, while for late O and early B stars many people prefer to use LTE models with LTE metal blanketing.

Concerning the non-LTE line formation problems of H and He II, Herrero (1987a, 1987b) has presented an improved treatment, based on the "accelerated lambda iteration method" developed by Werner and Husfeld (1985). This method permits, with decreased computational effort, to include more levels, more transitions, and also Stark broadening, in the statistical equilibrium equations. These additions produce noticeable improvements in the H absorption line profiles. The He II line profiles are also improved, but less. More important for He II lines is the recent work by Schöning and Butler (1988), who have calculated Stark broadening for He II using the "unified classical path theory" of Vidal et al. (1970). By comparing with the standard calculations (Auer and Mihalas, 1972), Schöning and Butler have found that the new theoretical profiles for the He II absorption line at 4686 Åhave weaker wings. Other He II diagnostic lines (4200, 4541) are not significantly affected.

All these recent results are leading to better spectroscopic determinations of T_{eff}, log g and He abundance for all kinds of hot stars. To give just one example, the new He II broadening theory has improved He abundance determination obtained from the spectra of several central stars of planetary nebulae (Mendez et al., 1988).

Work on other ions (including He I) is still largely based on the methods of Auer and Mihalas (1969) or Auer and Heasley (1976). Several groups have been actively working on non-LTE line formation calculations, using rather simple model atoms with few transitions. Sadakane (1986) gives a useful list of references. Some additional (or more recent) ones follow:

C II: Lennon, 1983;

C III: Sakhibullin et al., 1982;

C IV: Solov'eva, 1986;

N IV: Solov'eva, 1983;

O II: Brown et al., 1988;

Al III: Dufton et al., 1986;

Si II, III and IV: Lennon et al., 1986.

At the present time the Munich group (University Observatory) is carrying out an extensive application of the Auer and Heasley (1976) method to the non-LTE line formation problems of several ions: O II (Becker and Butler, 1988a), C II (Eber and Butler, 1988), N II (Becker and Butler, 1988b) C III, C IV, N III, N IV, N V, O III, O IV, O V, O VI, Si II, Si III, Si IV (in preparation). The main difference with previous work is a large increase in the number of energy levels and transitions used in each atomic model. The extra effort appears to be justified, at least in the case of C II. The complicated C II model used by Eber and Butler (1988) permits to reproduce the observed equivalent widths of the important C II line at 4267 Å, while the simpler model used by Lennon (1983) fails completely to do so.

What about the use of LTE line formation calculations for early B stars? Although this may be justified in some cases, the work by Becker and Butler (1988a,b) illustrates one advantage of non-LTE calculations: a smaller (and more reasonable) value of the microturbulence is enough to obtain a more or less horizontal plot of abundance versus equivalent widths of individual lines. Besides, the strongest lines of each ion are normally affected by stronger non-LTE effects. Therefore, LTE studies are more likely to produce systematic errors in the abundance determinations, particularly if just a few lines are used.

4. O-type stars on or near the main sequence

Two groups have made systematic quantitative spectroscopic studies of the earliest massive O-type stars: one using non-LTE, plane-parallel models without wind blanketing

(Kudritzki, 1980; Kudritzki et al., 1983; Simon et al., 1983), the other using wind-blanketed models (Bohannan et al., 1986; 1988). The two groups have only one object in common: ζ Puppis. For this star the agreement on the basic atmospheric parameters is quite good: T_{eff} = 42000 K, log g = 3.5, y = N(He)/(N(H) + N(He)) = 0.14 and 0.17. The corresponding uncertainties are 2000 K, 0.15 and 0.03. The unified model atmosphere (Gabler et al., 1988) yields a slightly higher gravity, but within the uncertainty. There seems to be no doubt that the He abundance in ζ Puppis is higher than normal (5 other massive early O stars show normal He abundance (y = 0.09), see the papers mentioned above). Of all these analyzed stars, ζ Puppis is also the one that lies farthest from the ZAMS. Therefore, the excess in He is interpreted as a consequence of evolution with mass loss and convective dredge-up (Maeder, 1983; 1987), which exposes material processed through the CNO cycle. If this interpretation is correct, then a high N abundance is expected in the atmosphere of ζ Puppis.

Butler and Simon (1985) have tried to test this prediction, using non-LTE line formation calculations for N III. Their analysis of the weak quartet lines (4514, etc.) led to a N abundance roughly 6 times the solar value, but they could not reproduce other, stronger N III lines in the spectrum of ζ Puppis, presumably because of wind effects. This indicates that reliable "metal" abundance determinations on the spectra of the earliest O-type stars are not possible yet.

The situation is better if we look at the late O stars. Schönberner et al. (1988) have made a careful non-LTE study of normal and N-rich late O and early B stars (the OBN stars, see Walborn, 1976). T_{eff}, log g and He abundance were determined, as for the earliest O stars, by fitting observed H and He absorption line profiles with theoretical non-LTE line profiles. The abundances of C, N, O were obtained by comparing the observed equivalent widths of lines of C II, C III, C IV, N II, N III, O II and O IV, in the optical and ultra- violet (IUE) spectra of the program stars, with the results of detailed non- LTE multilevel line formation calculations for all the relevant ions. The normal (comparison) stars yielded solar He, C, N and O abundances, with typical uncertainties of 0.02 in y and about 0.3 dex for CNO. The abundances in the atmospheres of the N-rich stars are quite different: N is enhanced (as expected) by at least one order of magnitude, C is depleted by about one order of magnitude, O is normal, and most interesting, He is noticeably enriched (y = 0.15 to 0.26). This He enrichment, which was not detectable with spectral classification techniques, is a strong indication that OBN stars are exposing CNO-processed material on their surfaces.

The positions of these N-rich stars in the log g - log T_{eff} diagram pose an interesting problem, because according to the standard evolutionary theory (Maeder, 1987) their surface abundances should be still normal. Schönberner et al. (1988) suggest that the missing ingredient may be enhanced mixing induced by fast rotation. See details in their paper.

5. The He abundance in h and χ Persei

Nissen (1974, 1976) has reported very low He abundances in the double cluster h and χ Persei, and in the association Cep OB III. If confirmed, this result would have important consequences for cosmological and stellar nucleosynthesis theories. His conclusions were based on differential studies of B stars, using photoelectric narrowband photometry of the He I line at 4026 Å, combined with photometric determinations of T_{eff} and log g.

Wolff and Heasley (1985) recalibrated Nissen's photometric index, using the results of their spectroscopic studies of He I 4026 in the spectra of main- sequence B stars. They suggest that the He abundance in Cep OB III may be normal, but support the low He/H ratio reported by Nissen for h and χ Persei, and conclude that narrow-band photometry is entirely adequate for the purpose of deriving He/H abundance ratios.

More recently, Lennon et al. (1988) have redetermined the He abundance in h and χ Per, using spectrograms of 7 stars. They obtain log g spectroscopically, by comparing observed and theoretical profiles for Balmer lines. Their gravities are systematically lower (by about 0.3 dex) than those determined photometrically by Nissen. Such low gravities are all that is needed to raise the He abundances to normal values. They conclude that the He abundance in h and χ Per is in good agreement with that found for other early-type, and suggest that the use of narrow-band photometry to derive He abundances in B stars may be compromised by the difficulty in determining reliable surface gravities.

For the moment, therefore, He abundances in the atmospheres of early-type stars cannot tell us much about the primordial He problem.

6. Interstellar abundance inhomogeneities?

Walborn (1976) has reported that all of the O9.5-BO.7 supergiants in the Orion Belt and in NGC 6231 (the nucleus of Scorpius OB1) have weak N lines relative to other supergiants with similar spectral types. This might be due to abundance anomalies in the progenitor interstellar material. Lennon and Dufton (1983) have tried to test this idea by studying 24 early-type main-sequence stars in NGC 6231. They determined abundances of He, C, N, O and Mg from moderate-dispersion spectrograms and both LTE and non-LTE calculations. All the elements turned out to have roughly normal abundances, except N, which was found to be deficient by about 0.3 dex. However, the N abundances were based on the equivalent widths of only one line, N II 3995. More recently, Brown et al. (1986) have presented a more extensive study of several southern clusters, based on similar methods, and have reported that N II 3995 gives too low N abundances. Having noted this effect, they reobserved two members of NGC 6231 at higher dispersions in several wavelength regions, and confirmed that the N abundances in NGC 6231, determined using 12 other N II lines, are normal.

According to this, now the "local" (a few Kpc) region of our Galaxy looks rather homogeneous, at least at the uncertainty level of about 0.3 dex, and the physical reason for the weakness of the N lines in the spectra of the supergiants in NGC 6231 remains unclear.

At the same time, Gehren et al. (1985) were suggesting that this simple and homogeneous picture may have to be modified. They have obtained high-resolution spectrograms of B-type main-sequence stars that are members of several Galactic open clusters at different galactocentric distances, in order to find independent evidence about abundance gradients (see next section). They claim a very high internal accuracy (0.1 dex) in the N and O abundances, because they work with many lines and use a strictly differential approach. Their abundances show a large scatter (at least 0.3 dex) from cluster to cluster, as well as within single clusters, and would therefore imply inhomogeneities everywhere and on a very small scale. However, they remark that their results should be taken as preliminary, and should be checked by enlarging their sample and by improving the model analyses, which were in LTE.

7. Abundance gradients in the Galaxy

The distribution of abundances in the interstellar medium is one of the fundamental observational constraints on theoretical models of the chemical evolution of our Galaxy. A great deal of effort has been devoted to obtain this information from the spectra of Galactic H II regions: see, e.g., Shaver et al. (1983). Nowadays this is accepted to be the most reliable method. However, the recent advent of echelle spectrographs with CCD detectors has made it possible to attempt an alternative approach: abundance determination using high-resolution spectrograms of B main-sequence stars in very distant open clusters.

The first preliminary results in this direction have been presented by Gehren et al. (1985) and Brown et al. (1986). Please refer to their papers for details. Figure 1 shows the O abundances obtained by Gehren et al. (1985), expressed as $12 + \log (O/H)$, as a function of the galactocentric distance in Kpc. For comparison, I have added the O abundances derived by Shaver et al. (1983). Figure 2 shows a similar plot for the N abundances. In both Figures I have also added abundance determinations by Brown et al. (1986) for the 3 open clusters they have in common with Gehren et al.: NGC 3293, NGC 4755 and NGC 6231.

In what follows I will disregard the possibility of small-scale inhomogeneities, raised by Gehren et al. (1985, see previous section), and will consider only the large-scale structure. From Figures 1 and 2 we can extract the following conclusions:

(a) In the case of O, the agreement with H II region abundances is tolerably good. Although Gehren et al. (1985) describe their results as discrepant with those of Shaver et al. (1983), in fact they can be reconciled if we interpret that the slope between 16 and 8 Kpc is almost zero (as suggested by Gehren et al.), and that the gradient becomes substantially steeper in the inner regions of the Galaxy, as already hinted by Shaver et al. A useful test would be to attempt a model atmosphere analysis of some B main-sequence stars 3 Kpc away in the direction of the galactic center, a task that, although not easy, may well be possible.

(b) In the case of N (Figure 2) the discrepancy is clear: the stellar N abundances are systematically larger. It is not clear which method of abundance determination should be blamed. Gehren et al. (1985) warn that their N abundances might be slightly overestimated because of their neglect of non-LTE effects. On the other hand, Pagel (1985) has mentioned several arguments that may be indicating a need to increase the N abundances derived from H II regions.

8. Accuracy of abundances

What conclusions can be extracted from this quick survey of the literature? Nowadays it is possible to obtain reliable He abundances, with typical uncertainties of 20%, or about 0.02 in y for normal abundances, from mid-B to the hottest known stars, above, on, or below the main sequence. The requisites are the following: for O-type stars, we must be able to make a simultaneous non-LTE spectroscopic determination of log g, T_{eff} and y. For B stars, at least log g must be determined spectroscopically. This level of accuracy

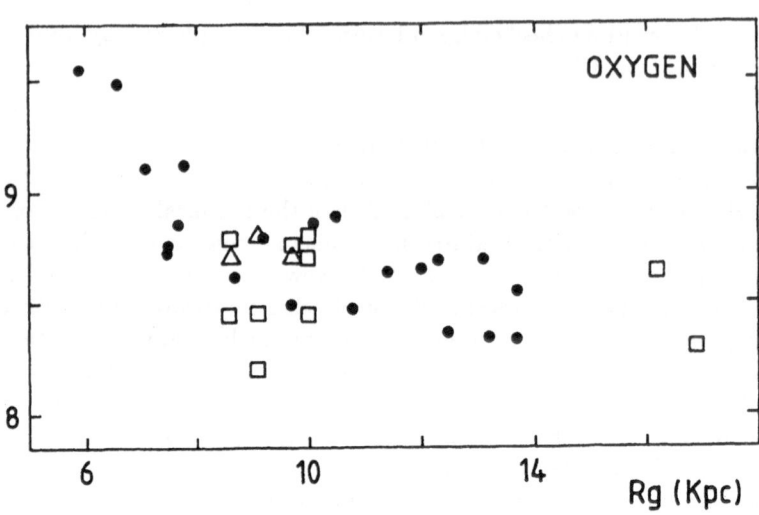

Figure 1. Oxygen abundances, expressed as $12 + \log (O/H)$, plotted against galactocentric distance. Filled circles are from Shaver et al. (1983). Squares are from Gehren et al. (1985). Triangles are from Brown et al. (1986).

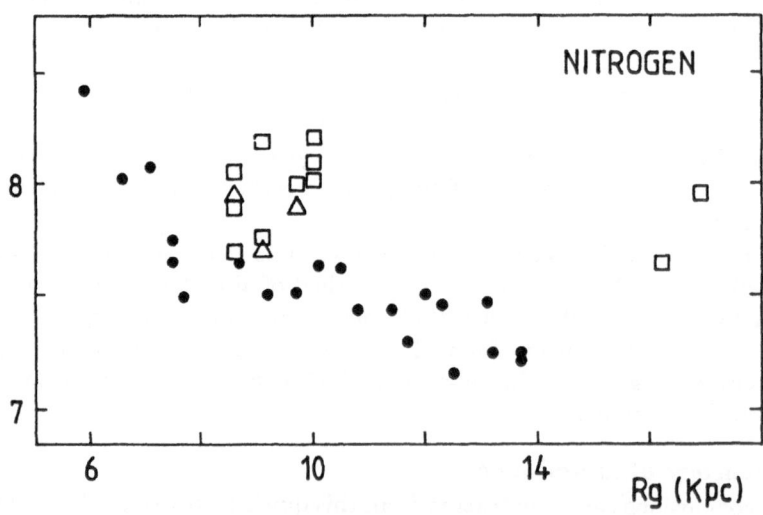

Figure 2. Nitrogen abundances, expressed as $12 + \log (N/H)$, plotted against galactocentric distance. Symbols as in Figure 1.

in the He abundances is sufficient to detect interesting evolutionary effects near the main sequence, and clearly has many uses in more advanced evolutionary stages.

Reliable "metal" abundances in the earliest O-type stars are not possible yet, but may be in the very near future. For the moment, we must content ourselves with order-of-magnitude estimates. As soon as non-LTE and wind effects become small (late O and early B stars) it is possible to obtain C, N, O, Mg, Al, Si abundances with uncertainties of about 0.3 dex. Here the requisites are: spectroscopic determination of gravity, and use of as many spectral lines of each ion as possible. Very careful and extensive non-LTE studies will be necessary in order to check if it is really possible to obtain, differentially, uncertainties as small as 0.1 dex.

Acknowledgement: I would like to thank K. Butler and R.P. Kudritzki (Munich University Observatory) for sending many preprints.

References:

Abbott, D.C. and Hummer, D.G. 1985, Astrophys. J. **294**, 286.

Abbott, D.C. and Lucy, L.B. 1985, Astrophys. J. **288**, 679.

Anderson, L.S. 1985, Astrophys. J. **298**, 848.

Auer, L.H. and Heasley, J.N. 1976, Astrophys. J. **205**, 165.

Auer, L.H. and Mihalas, D. 1969, Astrophys. J. **158**, 641.

Auer, L.H. and Mihalas, D. 1972, Astrophys. J. Suppl. **24**, 193.

Becker, S.R. and Butler, K. 1988a, Astron. Astrophys. **201**, 232.

Becker, S.R. and Butler, K. 1988b, Astron. Astrophys., in press.

Bohannan, B., Abbott, D.C., Voels, S.A. and Hummer, D.G. 1986, Astrophys. J. **308**, 728.

Bohannan, B., Voels, S.A., Abbott, D.C. and Hummer, D.G. 1988, in IAU Symposium 132, p. 127.

Brown, P.J.E., Dufton, P.L. and Lennon, D.J. 1988, Mon. Not. R. Astron. Soc. **230**, 443.

Brown, P.J.E., Dufton, P.L., Lennon, D.J. and Keenan, F.P. 1986, Mon. Not. R. Astron. Soc. **220**, 1003.

Butler, K. and Simon, K.P. 1985, Procs. of ESO Workshop on "Production and Distribution of CNO elements", p. 313.

Clegg, R.E.S. and Middlemass, D. 1987, Mon. Not. R. Astron. Soc. **228**, 759.

Dufton, P.L., Brown, P.J.F., Lennon, D.J. and Lynas-Gray, A.E. 1986, Mon. Not. R. Astron. Soc. **222**, 713.

Eber, F. and Butler, K. 1988, Astron. Astrophys., in press.

Gabler, R., Gabler, A., Kudritzki, R.P., Puls, J. and Pauldrach, A. 1988, Astron. Astrophys., submitted.

Gehren, T., Nissen, P.E., Kudritzki, R.P. and Butler, K, 1985, Procs. of ESO Workshop on "Production and Distribution of CNO elements", p. 171.

Herrero, A. 1987a, Astron. Astrophys. **171**, 189.

Herrero, A. 1987b, Astron. Astrophys. **186**, 231.

Kudritzki, R.P. 1976, Astron. Astrophys. **52**, 11.

Kudritzki, R.P. 1980, Astron. Astrophys. **85**, 174

Kudritzki, R.P. 1988, 18th Advanced Course, Swiss Soc. of Astron. and Astrophys., Leysin.

Kudritzki, R.P., Pauldrach, A. and Puls, J. 1987, Astron. Astrophys. **173**, 293.

Kudritzki, R.P., Simon, K.P. and Hamann, W.R. 1983, Astron. Astrophys. **118**, 245.

Lennon, D.J. 1983, Mon. Not. R. Astron. Soc. **205**, 829.

Lennon, D.J. and Dufton, P.L. 1983, Mon. Not. R. Astron. Soc. **203**, 443.

Lennon, D.J., Lynas-Gray, A.E., Brown, P.J.F. and Dufton, P.L. 1986, Mon. Not. R. Astron. Soc. **222**, 719.

Lennon, D.J., Brown, P.J.F. and Dufton, P.L. 1988, Astron. Astrophys. **195**, 208

Maeder, A. 1983, Astron. Astrophys. **120**, 113.

Maeder, A. 1987, Astron. Astrophys. **173**, 247.

Mendez, R.H., Kudritzki, R.P., Herrero, A., Husfeld, D. and Groth, H.G. 1988, Astron. Astrophys. **190**, 113.

Mihalas, D. 1972, NCAR, Boulder, Technical Note NCAR-TN STR-76.

Nissen, P.E. 1974, Astron. Astrophys. **36**, 57.

Nissen, P.E. 1976, Astron. Astrophys. **50**, 343.

Pagel, B.E.J. 1985, Procs. of ESO Workshop on "Production and Distribution of CNO elements", p. 155.

Pauldrach, A. 1987, Astron. Astrophy. **183**, 295.

Pauldrach, A., Puls, J. and Kudritzki, R.P. 1986, Astron. Astrophys. **164**, 86.

Puls, J. 1987, Astron. Astrophys. **184**, 227.

Sadakane, K. 1986, IAU Colloquium 90 (Upper main-sequence stars with anomalous abundances), eds. Cowley et al., Reidel, Astrophys. Space Sci. Library, **125**, 369.

Sakhibullin, N.A., Auer, L.H. and van der Hucht, K. 1982, Soviet Astron. **26**, 563.

Schönberner, D., Herrero, A., Becker, S., Eber, F., Butler, K., Kudritzki, R.P. and Simon, K.P. 1988, Astron. Astrophys. **197**, 209.

Schöning, T. and Butler, K. 1988, Astron. Astrophys., in press.

Shaver, P.A., McGee, R.X., Newton, L.M., Danks, A.C. and Pottasch, S.R. 1983, Mon. Not. R. Astron. Soc. **204**, 53.

Simon, K.P., Jonas, G., Kudritzki, R.P. and Rahe, J. 1983, Astron. Astrophys. **125**, 34.

Solov'eva, L.I. 1983, Soviet Astron. **27**, 415.

Solov'eva, L.I. 1986, Soviet Astron. **30**, 189.

Vidal, C.R., Cooper, J. and Smith, E.W. 1970, J. Quant. Spectr. Rad. Transfer **10**, 1011

Walborn, N.R. 1976, Astrophys. J. **205**, 419.

Werner, K. 1986, Astron. Astrophys. **161**, 177.

Werner, K. 1987, Ph. D. Thesis, University of Kiel.

Werner, K. and Husfeld, D. 1985, Astron. Astrophys. **148**, 417.

Wolff, S.C. and Husfeld, D. 1985, Astrophys. J. **292**, 589.

THE ACCURACY OF ABUNDANCES FROM MIDDLE B TO F MAIN SEQUENCE NORMAL STARS

Kozo Sadakane

Astronomical Institute, Osaka Kyoiku University

Tennoji-ku, Osaka, Japan 543

ABSTRACT: The accuracy of abundance determinations of iron in middle B to F type stars is reviewed based on recent results. The relative importance of various sources of the error contained in observational data and in physical data is discussed.

1. Introduction

The reliability of the results of stellar abundance determinations of main sequence middle B to early F-type stars has steadily improved in recent years. Improvements in the quality of observational data (both spectroscopic and photometric), in the reality of model atmospheres, and in the physical data such as transition probabilities and damping constants helped to increase the accuracy of the analysis. As an illustrative example, the historic change in the published abundance of iron (Fe) in the bright A0 main sequence star Vega is shown in Figure 1. In this figure, the abundances of iron relative to the sun are plotted against the year of publication between 1960 and 1986. Data before 1980 are taken from the compilation by Cayrel de Strobel (1985). Filled and open circles show data obtained from visual region and satellite UV region, respectively. Before 1980, the chemical composition of Vega had been believed to be identical with that in the sun. Now, published results of the abundance of iron in Vega converge around $\log \varepsilon(\text{Fe}) = -5.0$, and the deficiency of metals in this star seems to be established.

The accuracy of stellar abundance determinations depends on a combination of various factors. First, it depends on the human factor. When given the same observational material, different

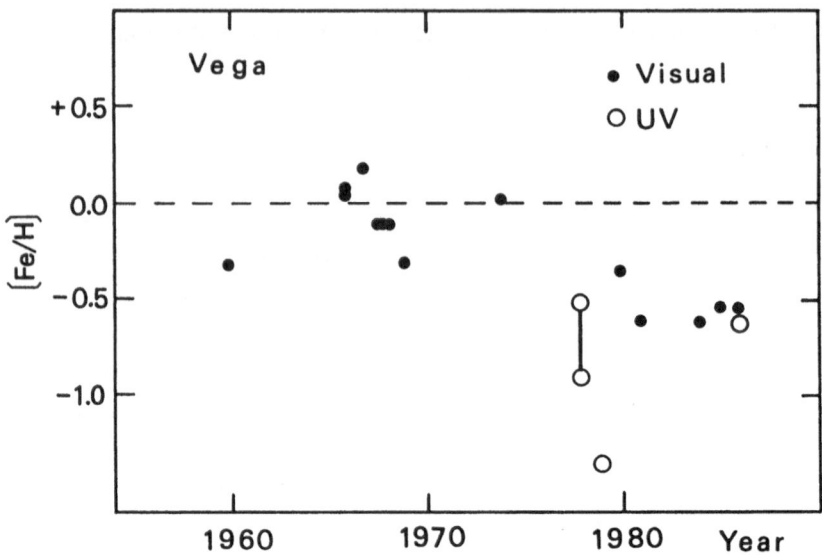

Fig. 1. Annual changes in the published abundances of Fe in Vega. The open circle at 1986 is taken from Sadakane et al. (1986).

person generally adopts different continuum levels, uses different methods of measurement, and selects different groups of lines
in his or her analysis. Second, it depends on the quality
(spectral resolution, S/N ratio, amount of scattered light, and
the calibration system) of the spectroscopic material. Further,
it depends on the adopted method of analysis (model atmospheres,
determinations of stellar parameters such as T_{eff}, log g, and the
microturbulent velocity). Finally, it depends on various
physical factors such as transition probabilities, damping
constants, and the assumed mechanism of line formation (LTE or
non-LTE). Gustafsson (1988) discussed the effects of some of
these factors on stellar abundances for various types of stars.

I will discuss which of the above factors are dominant in
determining the accuracy of abundance determinations of B, A and
F type stars. The discussion is based on a collection of
published data after 1980. I will concentrate here on the
abundances of iron, since the element is studied more throughly
than any other elements in these stars.

2. The data

TABLE 1

Abundance determinations of iron in normal stars

Star	HR	Sp.T.	T_{eff}	log g	ξt	Fe I	N	Fe II	N	Ref.
α Lyr	7001	A0V	9650	3.90	2.5[a]	-4.5[b]	14	-4.9	13	9
			9660	3.94	2.0	-5.09	28	-5.09	23	16
			9500	3.9	1.5-1.9	-5.02	14	-5.10	13	13
			9650	3.9	2.0	-4.89	40	-4.85	37	18
			9500	3.9	2.0[c]	-4.76[b]	35	-5.01[b]	45	10
θ Leo	4359	A2V	9400	3.60	2.5	-4.48	32	-4.40	11	17
			9300	3.4	2.5	-4.42	23	-4.51	19	12
			9250	3.55	1.7	-4.51	163	-4.39	64	4
η Lep	2085	F1III	7000	4.1	2.3	-4.55	17	-4.71	8	14
			6850	4.05	3.4	-4.88	103	-4.78	17	3
α CMi	2943	F5IV-V	6650	4.0	1.8	(-0.01)[d]	66	(+0.03)[d]	21	11
			6750	4.04	2.1	(-0.01)[d]	94	(+0.06)[d]	16	19
			6400	3.95	1.8	-4.44	12	-4.06	7	14
τ Her	6092	B5IV	14750	3.75	0.0	——	—	-4.80	29	4
	2154	B5IV	14500	3.12	1.0	——	—	-4.75	11	1,6
	5780	B6IV	14000	4.0	0.7	——	—	-4.45	21	1,6
π Cet	811	B7V	13150	3.65	0.0	——	—	-4.42	47	1,6
21 Aql	7287	B8III	13000	3.3	0.2	——	—	-4.40	49	1,6
5 Aqr	7985	B9III	11150	3.35	0.7	-4.36	10	-4.18	44	2
14 Cyg	7483	B9III	10850	3.45	1.4	-4.21	10	-4.36	34	2
134 Tau	2010	B9IV	10825	3.9	1.6	-4.28	29	-4.47	60	1,6
θ Aql	7710	B9.5III	10375	3.3	1.5	-4.5	20	-4.9	36	1,6
υ Cap	7773	B9.5V	10250	3.9	1.6	-4.40	38	-4.28	62	7
21 Peg	8404	B9.5V	10500	3.5	1.0	-4.76	32	-4.75	23	15
	7338	A0III	9500	3.0	0.5	-5.33	25	-5.28	22	15
α Dra	5291	A0III	10075	3.30	0.4	-4.89	44	-4.90	48	5
τ Gem	2421	A0IV	9300	3.40	2.0	-4.45	123	-4.43	45	18
θ Vir	4963	A1IV	9500	3.6	1.5	-4.34	39	-4.37	26	8
η Oph	6378	A2V	9300	4.20	3.0	-4.54	48	-4.59	32	17
	6559	A7IV	7900	3.2	3.3	-4.66	117	-4.73	24	3

[a] Doppler broadening velocity (DBV).   [c] Depth-dependent.

[b] non-LTE result.   [d] Determined relative to the Sun.

References to TABLE 1:
(1) Adelman 1984, (2) Adelman 1986, (3) Adelman 1987, (4) Adelman
1988, (5) Adelman et al. 1987, (6) Adelman and Fuhr 1985, (7)
Adelman and Nasson 1980, (8) Dobrichev et al. 1987, (9) Dreiling
and Bell 1980, (10) Gigas 1986, (11) Kato and Sadakane 1982, (12)
Klochkova et al. 1985, (13) Lane and Lester 1984, (14) Lane and
Lester 1987, (15) Sadakane 1981, (16) Sadakane and Nishimura
1981, (17) Savanov 1985, (18) Savanov and Khalilov 1985, (19)
Steffen 1985.

   Data of abundances of iron in B, A, and F type stars (spectral
type between B5 and F5) published after 1980 are collected.
Main sequence superficially normal stars and several luminosity
class III stars (giants) are included. Only those results which
are based on high dispersion data obtained in the photographic
region are collected. Model atmospheres are used in all of the
cases. Relevant data are given in Table 1. Star names, spectral
types, adopted stellar parameters, obtained abundances of iron
from Fe I and Fe II lines and the number of used lines are given.
References are given in the footnote. When multiple analyses are
carried out for a star by the same worker, only the latest result
is listed in the Table. Averaged abundances of iron relative to
the sun in these 21 stars are plotted against the effective
temperature in Figure 2. When multiple entries are given for a
star in Table 1, the star is shown on Figure 2 by a rectangular
box which covers the range of published effective temperatures
and abundances. Sizes of these boxes give an indication of the
present day accuracy of abundance determinations. We can find no
star which shows a definite overabundance of iron in the present
sample. On the other hand, deficiencies of iron are found in
several stars including Vega.

2.a. The case of Vega (α Lyr)

   Five independent analyses of Vega have been published since
1980. Table 2 shows technical details of these analyses.
Spectroscopic observations were carried out at four obser-
vatories. Photographic plates were used in three analyses and
a solid state detector (reticon) was used in the remaining two.

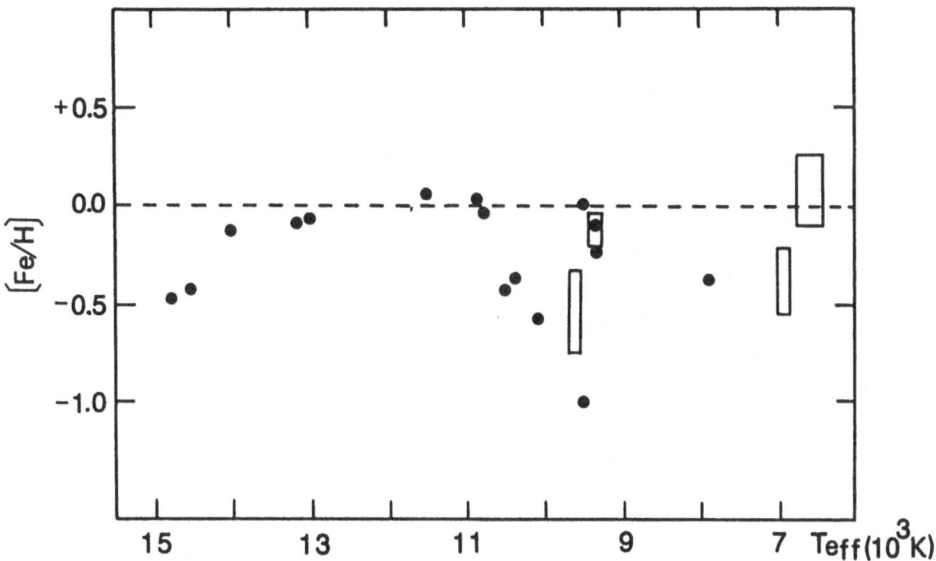

Fig. 2. Published abundances of Fe in 21 superficially normal
stars given in Table 1. Rectangular boxes show the range in
published T$_{eff}$ and abundances for the four stars (α Lyr, θ Leo,
η Lep, and α CMi).

Lines are selected using different criteria in each analysis and
the adopted log gf values are not identical. Dreiling and Bell
(1980) applied a correction for the non-LTE effect. Gigas (1986)
carried out detailed non-LTE analyses for both Fe I and Fe II and
found that effects of non-LTE on Fe II lines are very small. The
LTE line formation is assumed in the remaining three studies. In
spite of these differences in observational data and in the
method of analysis, resulting abundances of iron from Fe II lines
agree within +0.15 dex. A good agrrement is also found in the
derived microturbulent velocities. These examples show that a
careful analysis based on high quality data results in a fairly
reliable abundance.

2.b. τ Her and α Dra

There are several stars for which multiple analyses are carried
out by a same person (or a group) based on different observational data. Technical details of two such examples (τ Her and
α Dra) are summarized in Tables 3a and 3b. Adelman (1988) showed

TABLE 2

Analyses of Vega (A0 V) by various authors

based on different observational data

Author	(1)	(2)	(3)	(4)	(5)
Observatory	McDonald	Okayama	McDonald	Crimea	Mt. Wilson
Detector	Reticon	Photogr.	Reticon	Photogr.	Photogr.
Resl.(Å/mm)	0.2-0.3 Å	4.1	0.2-0.3 Å	1.5-3.0	1.1
T_{eff}(K)	9650	9660	9500	9650	9500
log g	3.9	3.94	3.9	3.9	3.9
ξt (km/s)	2.5(DBV)	2.0	1.5-2.0	2.0	2.0
log ε(Fe I)	-4.5*	-5.09	-5.02	-4.89	-4.76*
N	14	28	14	40	35
log ε(Fe II)	-4.9	-5.09	-5.10	-4.85	-5.01*
N	13	23	13	37	45

* non-LTE

References: (1) Dreiling and Bell 1980, (2) Sadakane and
Nishimura 1981, (3) Lane and Lester 1984, (4) Savanov and
Khalilov 1985, and (5) Gigas 1986.

TABLE 3a

Analyses of τ Her (B5 IV) based on different data

Author	Adelman (1986)	Adelman (1988)
Observatory	Mt. Wilson, KPNO	DAO
dispersion(Å/mm)	4.3 - 15.6, 8.9	2.4
T.ₜₜ(K)	14750	14750
log g	3.65	3.75
₤t(km/s)	1.8	0.0
log ₤(Fe II)	-4.47	-4.80
Number of lines	12	29
r.m.s.	0.23	0.20

$$W\lambda \ (DAO) = \underline{0.870} * W\lambda \ (Mt. \ Wilson) \ -6.27 \ (m\text{Å})$$

TABLE 3b

Analyses of α Dra (A0 III) based on different data

Author	Adelman et al.(1987)	Adelman et al.(1987)
Observatory	KPNO	DAO
Dispersion(Å/mm)	8.9	2.4
T.ₜₜ(K)	10075	10075
log g	3.30	3.30
₤t(km/s)	0.8	0.4
log ₤(Fe I)	-4.69	-4.89
Number of lines	17	44
r.m.s.	0.12	0.27
log ₤(Fe II)	-4.44	-4.90
Number of lines	25	48
r.m.s.	0.16	0.21

$$W\lambda \ (DAO) = \underline{0.680} * W\lambda \ (KPNO) \ - \ 0.00 \ (m\text{Å})$$

that equivalent widths of lines in τ Her measured using DAO
plates (2.4 Å/mm) are systematically smaller than those measured
using Mt. Wilson plates of lower dispersion (Figure 3). This
difference resulted in a remarkable revision of the microturbulent velocity in τ Her from 1.8 km/s to 0.0 km/s. Also, the

abundance of iron obtained from Fe II lines reduced by 0.38 dex
(by a facor of 2.5!). Adelman et al. (1987) found a significant
difference in the measured profile of Hᴦ of α Dra between DAO
plates and KPNO plates. The difference affects the determination
of the stellar surface gravity. They preffered to use the
profile obtained from the DAO plates. They obtained a lower
abundance of iron and a lower microturbulent velocity from DAO
plates than from KPNO plates in α Dra. These examples show that
results of stellar abundance analysis depend on the dispersion of
the used spectroscopic material and/or on the used calibration
system.

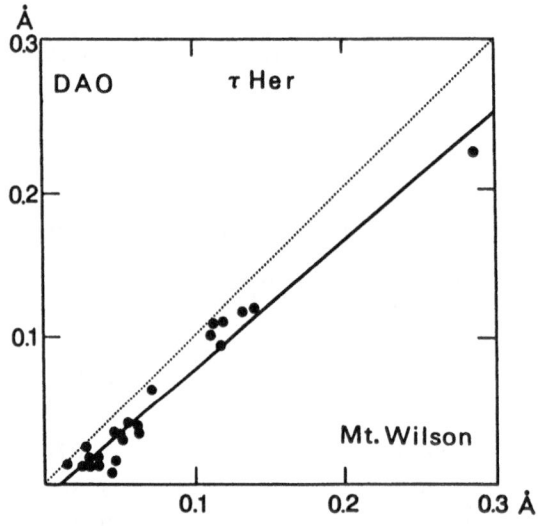

Fig. 3. Comparison of measured equivalent widths in τ Her from
DAO and Mt. Wilson plates. The solid line shows the relation
adopted by Adelman (1988).

2.c. The case of Procyon (α CMi)

Kato and Sadakane (1982) and Steffen (1985) carried out
analyses of Procyon using the Griffin's (1979) spectroscopic
atlas. Both of these analyses were done differentially relative
to the sun. Thus, their results are free from errors in the log
gf values. As we can see on Table 1, resulting abunadnces of
iron in Procyon agree quite well each other and their results

TABLE 4

Comparison of the r.m.s. scatter (Procyon)

Ion	KS[a] r.m.s.	N	ST[b] r.m.s.	N	LL[c] r.m.s.	N
Ti II	0.08	15	0.10	14	0.31	7
Cr I	0.08	30	0.07	30	0.23	5
Cr II	0.10	17	0.09	17	0.26	6
Mn I	0.10	18	0.10	15	0.25	8
Fe I	0.09	66	0.09	94	0.24	12
Fe II	0.08	21	0.07	16	0.22	7

[a]Koto and Sadakane (1982), [b]Steffen (1985), [c]Lane and Lester(1987).

show that the abundance of iron in Procyon is very close to the solar value. Lane and Lester (1987) published abundance data of classical Am stars. They used Procyon as a standard star. Their study is based on spectroscopic materials obtained at the KPNO. They obtained a similar microturbulent velocity in Procyon as found by Kato and Sadakane (1982) or by Steffen (1985). On the other hand, they obtained a higher abundance of iron from Fe II lines (log ε(Fe) = -4.06) using published log gf values. We compare in Table 4 r.m.s. scatters in their results with those in Kato and Sadakane (1982) and in Steffen (1985) and found that the scatter in Lane and Lester (1987) is twice as large. This difference most probably reflects errors in the log gf values. Thus, even the most recent data of log gf values used by them are not free from random errors.

3. Error analysis

Gigas (1986) presented quantitative estimations of errors which are introduced by various sources in the abundance of iron obtained from both Fe I and Fe II lines for Vega. Besides the difference between LTE and non-LTE, the primary sources of uncertainty in the abundance obtained from Fe I lines are errors in T_{eff} and in electron collision cross sections. The secondary sources are errors in log g and in photoionization cross sections. On the other hand, for Fe II, the primary source is the error in log gf values and the secondary one is the error in the microturbulent velocity. The total estimated uncertainty in the obtained abundance is $\pm$ 0.22 dex for both Fe I and Fe II. Lane and Lester (1987) tabulated estimations of errors which are caused by changes in T_{eff}, log g, and in microturbulent velocity for various ions using model atmospheres of F-type stars. Combining these results, we conclude that the expected error in the results of abundance determinations of iron is not smaller than $\pm$ 0.2 dex even in the case of Vega. Expected errors are generally larger than $\pm$ 0.3 dex for most of the stars listed in Table 1. Errors in the results of other elements may be larger than this because of the uncertainty in log gf values and also in other parameters such as the damping constants.

4. Summary

We have discussed various sources of the error which introduce uncertainty in the stellar abundance determinations. The analysis of error sources is not straightforward, because different sources become dominant in different kind of stars or for different kind of ions. The effect of various sources of error also depends on the strengths of the used lines. For example, the uncertainty in the damping constants and that in the microturbulent velocity affect most seriously the results from strong and intermediately strong lines, respectively. A good S/N ratio is critical in obtaining reliable abundances from weak lines. However, a general conclusion is that the quality (resolution and S/N ratio) is critically important in stellar abundance studies. It is now highly desirable to establish high quality spectroscopic standards for various types of stars which replace the classical work by Wright et al. (1964). Another general conclusion is the need to improve the quality of physical data, especially that of log gf values.

This work was supported in part by a Grant-in-Aid for Scientific Research of the Japanese Ministry of Education, Science, and Culture (61540184).

REFERENCES

Adelman, S.J. 1984, M.N.R.A.S., 206, 637.
————. 1986, Astr. Ap. Suppl., 64, 173.
————. 1987, Astr. Ap. Suppl., 67, 353.
————. 1988, M.N.R.A.S., 230, 671.
Adelman, S.J., Bolcal, C., Kocer, D., and Inelman, E. 1987, P.A.S. Pacific, 99, 130.
Adelman, S.J., and Fuhr, J.R. 1985, Astr. Ap., 152, 434.
Adelman, S.J., and Nasson, M.A. 1980, P.A.S. Pacific, 92, 346.
Cayrel de Strobel, G. 1985, in IAU Symposium 111, Calibration of Fundamental Stellar Quantities, ed. D.S. Hayes, L.E. Pasinetti, and A.G.D. Philip (Dordrecht: Reidel), p. 137.
Dobrichev, V.M., Ryabchikova, T.A., and Raikova, D.V. 1987, Astrophysics, 26, 31.
Dreiling, L.A., and Bell, R.A. 1980, Ap. J., 241, 736.
Gigas, D. 1986. Astr. Ap., 165, 170.
Griffin, R. & R. 1979, A Photometric Atlas of the Spectrum of Procyon, Cambridge, England.
Gustafsson, B. 1988, in IAU Symposium 132, The Impact of Very High S/N Spectroscopy on Stellar Physics, ed. G. Cayrel de Strobel, and M. Spite (Dordrecht: Kluwer), p. 333.
Kato, K., and Sadakane, K. 1982, Astr. Ap., 113, 135.
Klochkova, V.G., Panchuk, V.E., and Tsymbal, V.V. 1985, Bull. Spec. Astrophys. Obs. North Caucasus, 19, 19.
Lane, M.C., and Lester, J.B. 1984, Ap. J., 281, 723.
————. 1987, Ap. J. Suppl., 65, 137.
Sadakane, K. 1981, P.A.S. Pacific, 93, 587.
Sadakane, K., and Nishimura, M. 1981, P.A.S. Japan, 33, 189.
Sadakane, K., Nishimura, M., and Hirata, R. 1986, P.A.S. Japan, 38, 215.
Savanov, I.S. 1985, Izv. Krymskoj Astrofiz. Obs., 73, 92.
Savanov, I.S., and Khalilov, A.M., 1985, Izv. Krymskoj Astrofiz. Obs., 72, 106.
Steffen, M. 1985, Astr. Ap. Suppl., 59, 403.
Wright, K.O., Lee, E.K., Jacobson, T.V., and Greenstein, J.L. 1964, Publ. Dominion Astrophys. Obs., 12, 173.

ACCURACY OF THE DETERMINATION OF THE ABUNDANCES IN SOLAR TYPE STARS

F. Spite
Observatoire de Paris, Section de Meudon
92195 Meudon CEDEX, France

1. Introduction

The accuracy of abundance determination is limited, as usual, by two kinds of errors : the random errors and the systematic errors. The random errors are rather easily evaluated, since the abundance of an element may be derived from the analysis of any suitable line of this element : the scatter of the values, derived from individual lines, provides some information about the random error. The systematic error is much more difficult to estimate. I will try to do it, limiting the exercise to the dwarf stars. Of course the separation between random and systematic errors is in fact somewhat artificial. I will however keep it here. I will on many points, rely on papers by Gustafsson (mainly : Gustafsson 1988).

2. The measurement errors

Nowadays, we are in a favorable position about the random errors. Modern detectors have brought a significant improvement about the accuracy of the measurements. For the Sun, which is the fundamental standard for the solar type stars, the use of a photoelectric detector permitted the edition of a superb solar atlas (Delbouille et al. 1973) completed by atlases of the solar flux (Beckers et al. 1976, Kurucz et al. 1984). The solar atlases are extended in the UV and in the IR. A particularly interesting document is the atlas showing the fitting of a synthetic and an observed solar spectrum in the UV (Kurucz and Avrett 1981).

For the stars, the advent of solid state detectors has brought a definite improvement in the analysis of all type of stars, and a recent IAU Symposium (G. Cayrel and M. Spite, eds, 1988) has illustrated this point; but the improvement has probably been greater for the solar type stars than for hot stars. The spectra of the solar type stars are crowded with overlapping lines and bands in the blue region and have relatively undisturbed regions only in the red part of the spectrum; the new detectors are especially sensitive in this red region. The new detectors brought the possibility to obtain spectra with a higher signal-to -noise ratio and/ or a better resolution (which helps to avoid the instrumental blending of lines). The response of the solid state detectors is linear (outside the saturation region) and brings a higher photometric accuracy.

Of course the new detectors have also some drawbacks : they have parasitic signals in the form of uneven dark current and spikes (due to the so-called cosmic rays); they have fringes, which are not easy to correct for by using "flat field" exposures (because the beam of the lamp does not, in general, illuminate the chip in exactly the same way as the stellar beam). Some of the chips have a persistence effect. Some have a low transfer efficiency, which demands a pre-flashing of the chip, and this process adds some extra noise. The unequal sensitivity of the pixels is taken into account, dividing by flat field exposures (such a process brings some additionnal uncertainty). The bad columns can be taken care of, although a lost information cannot be entirely recovered.

Another important drawback of the CCD detectors is that the chips are up to now of limited size, which means a limited spectral range per exposure.

Therefore the number of the measured lines is generally smaller than the number obtained when using other detectors such as the photographic plates : the greater accuracy of the measurements compensates for the smaller number of the lines. Sometimes an échelle configuration is used in order to alleviate this problem, which will disappear when larger chips, or a mosaic of juxtaposable chips, will be available.

All in all, the new detectors are able to provide rather easily, more accurate data (fig. 1). I cannot omit, however, the excellent work made by R. F. Griffin and R. Griffin (1968, 1979) using photographic plates (Arcturus and Procyon atlases) ; also the data obtained by Arpigny and Magain (1983) on metal-poor dwarfs by co-adding numerous photographic spectra. But such achievements are rather exceptionnal, since few people had the patience to slowly accumulate a number of high quality spectra ! Of course not only patience but numerous hours of telescope are necessary.

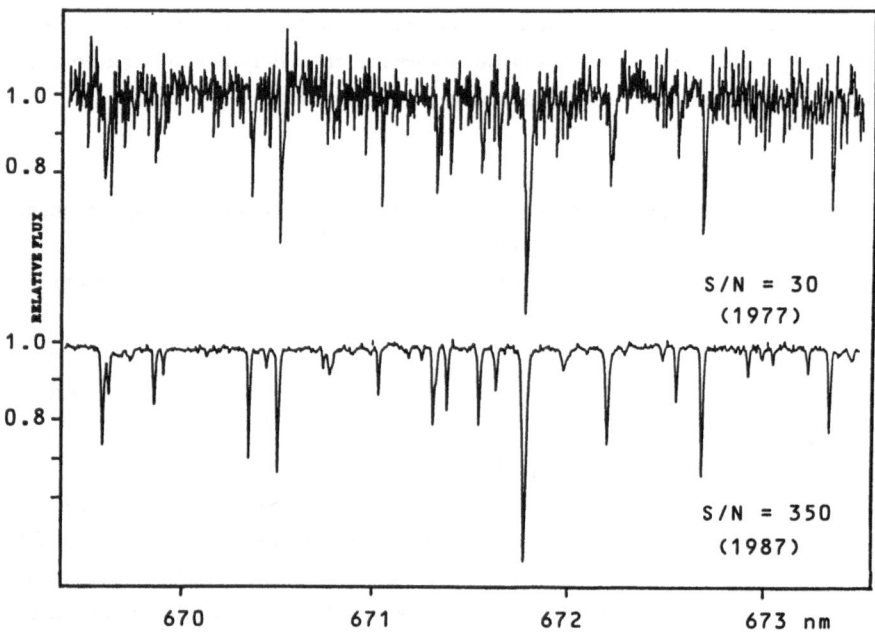

Fig. 1. The star HD 76151 is observed with a higher signal to noise ratio when using a modern detector : this observational progress brings in turn interesting advances (see IAU Symposium 132, G. Cayrel de Strobel and M. Spite, eds., 1988).

Let us point out that an increased accuracy in the measurement of the spectral lines will bring only a limited improvement of the abundance accuracy for the elements which have a large number of lines (such as iron for example). One of the improvements due to the new detectors is the possibility to measure faint lines, i. e. the lines which really define the abundance : we will come back to this point about the determination of the microturbulence parameter. The real improvement is for the elements which show only a few lines in the spectrum : here the measurement accuracy is important and the new detectors permit such an accuracy.

Finally the random errors may be reduced at present time, owing to the technical progresses made about the detectors. One of the interest of better observational data is that they will undoubtedly provide a more critical comparison between models and observations, showing how adequate are the models and how to improve them.

3. The systematic errors

From where are coming the systematic errors ?

In principle, for normal stars, the method of determination of the abundances, using an observed spectrum, is simple. Synthetic spectra are calculated from a multi dimensionnal grid of theoretical models, each defined by several values of the basic parameters : the best fit between the observed and the calculated spectrum defines the model which will be finally retained, and its basic parameters. The uncertainty in this process (systematic errors) arise essentially from the fact that the theoretical model is too simple and is only an approximation of the real star; moreover the optimum fitting is not always well defined : if the models of the grid are not adequate, *some* features may be fitting well the observations and *some* other not. The situation is basically the same if the theoretical models are replaced by a set of empirical models, built with well defined recipes. In fact, for several reasons, the method used in practice is the successive determination of the parameters T_{eff}, log g of the model, followed by the line analysis, which provide the microturbulence parameter and the abundances (see the appendix).

3.1 Observed indices of the inadequacy of the model atmospheres

Are the generally used models suspected of bringing systematic errors ?

Up to now, the available models for normal stars are in general simple, i. e. plane parallel layers, hydrostatic equilibrium, radiative equilibrium with blanketing and in LTE, the convective flux is computed by using the mixing length approximation. These "theoretical" models are defined by only a few parameters: the effective temperature, the gravity, the metallicity, and the microturbulence parameter; for the adjustment of profiles, the macroturbulence and the rotation parameter *v sini* have to be determined, although these parameters do not change the equivalent widths of the lines. In practice, it is generally not possible to disentangle the rotation and the macroturbulence. In principle the rotation of the star should be taken into account in the model.

In fact it is well known from the analysis of the solar spectrum that such models are only approximations of the reality. The photosphere of the Sun shows spots, granules etc.... However, the exact amplitude of the errors induced by these simplifications are not yet well known. Computations have been made by Dravins (1988) which represent the solar case. The error introduced is not easy to evaluate even for the Sun, and the amplitude of the phenomenon on other stars is unknown.

It has been noted, a long time ago, that some segregation occurs between the lines of zero eV excitation level, and the other lines (of higher levels) of the same element. The computed UV spectrum of the Sun does not fit well (fig. 2) the observed spectrum (Kurucz and Avrett, 1981). More important : the colors predicted by the models represent the real colors only after a zero-point correction, which is not yet well defined and is subject to controversy (we will come back to this topic). Magain (1985) claims that the models of Gustafsson et al.(see Gustafsson et al. 1975 and references therein) are not able to represent both the continuum and the line spectrum (including the profiles of the Balmer lines) for two halo dwarfs. Steffen (1985), analysing the nearby star Procyon (α CMi) finds that neither a theoretical nor an empirical model can describe entirely the observed spectrum, and he had to increase the initial value of T_{eff} (found from analysis of the flux in the continuum) for his abundance analysis. In other words, there is a discrepancy between the temperature deduced from the color and the temperature deduced from the ionization equilibrium (see also the communication by Dr. Sadakane in this volume). Let us note that the empirical and theoretical solar models show definite temperature differences (at most 200K in some layers). Finally Gustafsson notes that the comparison between models based on similar assumptions show sometimes temperature differences up to 100K: it seems that these differences have to be attributed mainly to differences in the physical data used in the computations.

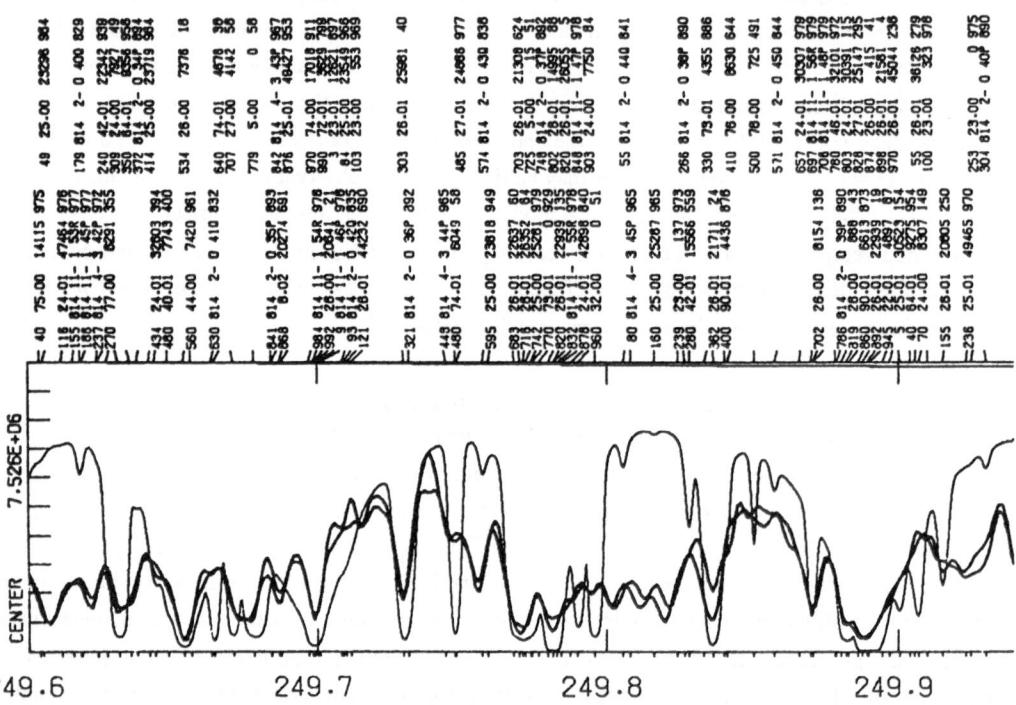

Fig. 2. *Synthetic spectra (thin curve) in the UV do not yet match the observed spectra (thick curve; Kurucz and Avrett, 1981). The numbers in the upper panel indicate the spectral lines that contribute to the energy distribution.*

3.2 A particular aspect of the problem : the colors

The cool dwarfs have the great advantage of being similar to the only well known star : the Sun. The models for cool dwarfs, inspired from the Solar models, are especially reliable. Therefore, the situation should be excellent for the cool dwarfs. It is not yet entirely so.

A completely differential analysis of a star relative to the Sun would require a fit of the colors. Unfortunately, the direct measurements of the colors of the Sun are extremely difficult, ranging (see Saxner and Hammarbäck,1985) from $B-V = 0.66$ (Kron, 1963) to $B-V = 0.68$ (Gallouët, 1964). Indirect determinations of the solar colors have met with some difficulties, mainly because it is not so easy to find a star identical to the Sun. It appears that it is not enough to have similar temperature, gravity and metallicity. The microturbulence is an important factor for the color (Conti and Deutsch 1966) but also the chromospheric activity of the star may have an influence (Campbell, 1984, LaBonte and Rose, 1985).

Abundance anomalies may also be an important factor. Helium abundance (Nissen 1988) has often be quoted : this parameter is not accessible to direct observational checks in cool stars. Other abundance anomalies could also be responsible for color variations. The rotation of the stars has an influence on the color through the modification of the stellar structure. The number of the possible factors explain why the problem of the color of the Sun is so difficult to solve.

From the measurement of the wings of Balmer lines from photographic spectra, Gehren (1981) finds some deviation from the the color temperature relations derived from the models.

In spite of these difficulties, the colors have been calibrated and provide one of the best way to determine the effective temperature of the cool stars. In the method of Blackwell et al. (1977, 1986) and its application by Saxner and Hammarbäck, the calibration is hardly model dependent (we will come back to this point). Then we can estimate the temperature error. The slope of the calibration for solar type stars is such that for $\Delta(B-V) = 0.01\,mag$, the temperature error is about 40K, and the standard deviation of a good (homogeneous) photometry is about 0.01mag.

The determination of the zero point of the colors of the models is made by several methods. Johnson (1966) made a calibration of colors, using measurements of the angular diameter of the stars. Popper (1980) published a calibration based on eclipsing binaries. Other observations such as lunar occultations, interferometry ...will improve the calibrations in the future.

The colors permit an useful check : stars of similar T_{eff}, log g , metallicity and microturbulence should have exactly the same color. This is not obvious from the observations, so that at least one additional parameter should be at work (see Alexander 1986, Nissen 1988).

3.3 A word of caution : are random and systematic errors always clearly separated ?

The difference is not so sharp as it could be thought at first.

For example each line of an element provides an individual value of the abundance. The scatter of these values is attributed in part to the measurement errors (either the equivalent widths or the adjustement of profiles). With more accurate data, the scatter decreases indeed in most stars, reflecting the greater accuracy of the measurements (fig. 3). As an example of an extreme case, let us note that the scatter does not decrease as much as expected (fig. 4) for the supergiant stars. This could mean that the model does not take into account one or several phenomena which affect the individual lines.

Which are these phenomena, and how to take them into account is another problem, obviously quite difficult (otherwise, it would have already been attacked !). No completely defined and self consistent theory of the hydrodynamics of stellar atmospheres is available at present; so that any attempt to include such phenomena in the models, not only increases severely the difficulty of the task of the modelists, but also introduces an additional number of free parameters and nobody knows how to cope reasonably with such a situation.

3.4 Check of the adequacy of the models

It is important to know how much the presently available models may be inadequate. Let us examine a few possible checks of the models.

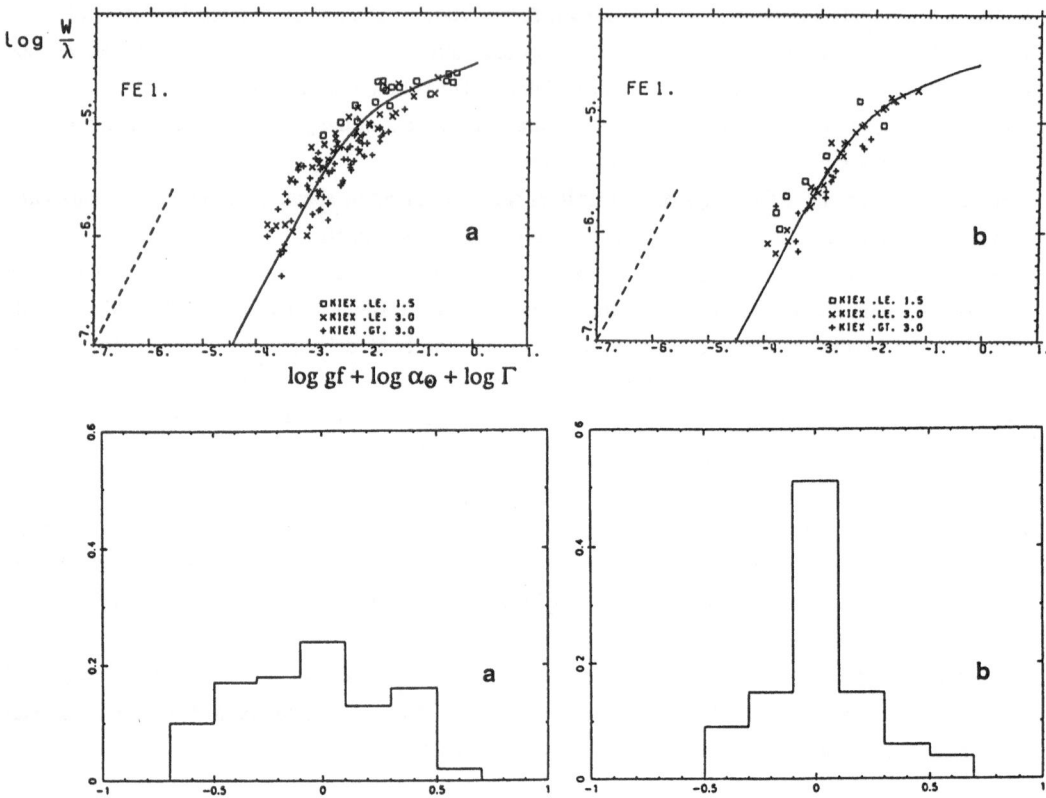

Fig. 3   Upper part : curves of growth of the star HD 184711 derived from spectra obtained with photographic plates (a) and with a CCD (b) respectively. The abscissae are those defined by Cayrel and Jugaku. The scatter is larger for the photographic measurements. Lower part : the histograms show quantitatively the difference of scatter. The abscissae are the deviations (in logarithmic units) from the mean abundance.The modern detectors enable a decrease of the random errors .

3.4.1 Check of the Solar models

The theoretical solar models are not entirelly satisfactory. The HM empirical model (Holweger and Müller,1974) reproduces solar observations much better than the Kurucz model (Steffen, 1985 and references therein). Since much more information is available for the Sun than for any other star (except the colors) the situation may look uncomfortable. Fortunately, the discrepancies encountered between observed and computed spectra are relatively small.

For stellar models, let us examine what is the result of the comparison of independent determinations of one of the basic parameters.

3.4.2 Determinations of T_{eff}

The effective temperature T_{eff} is usually determined from colors, Balmer lines profiles and excitation equilibrium.

  Here are evaluations of the errors in the derivation of T_{eff} from the color, using the calibrations of

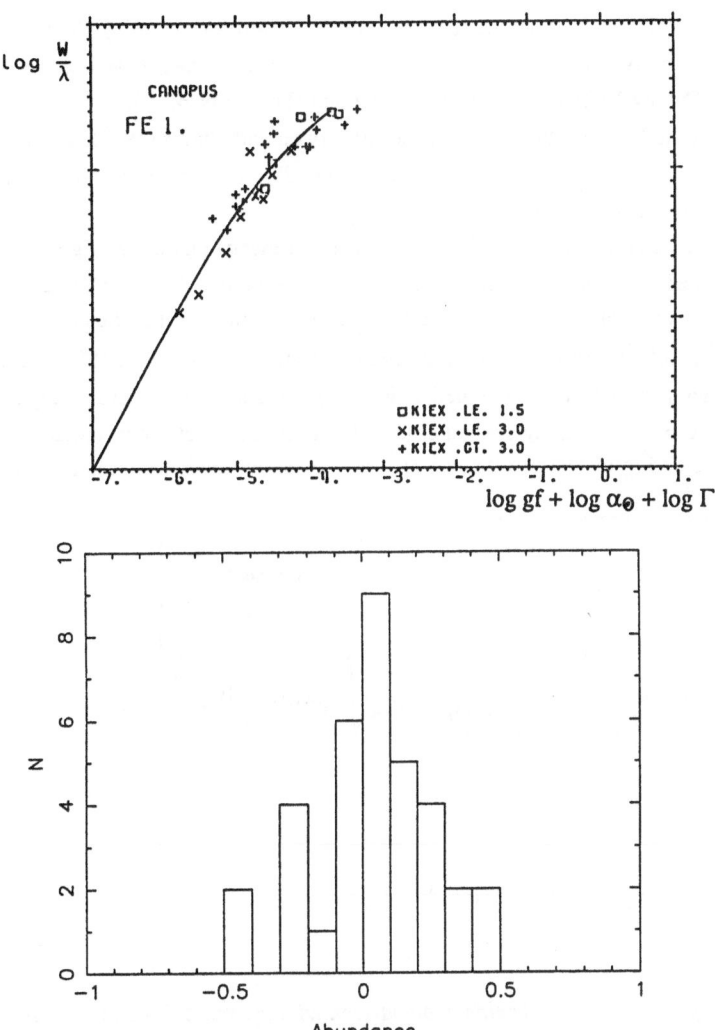

Fig. 4. Observations of lines in supergiants, even using modern detectors and high signal to noise ratios, remain scattered, in spite of the reduced random errors.

Johnson (1966) or Popper (1980). Color indices such as *R-I*, *V-R* and *V-I* vary rapidly with the temperature : this is why they are widely used for temperature determination. Moreover the color index *R-I* is found to be largely independent of the metallicity.

For solar type stars, the temperature increases by 50K when the *R-I* color decreases by 0.01 magnitude. The accuracy of the color index is about 0.01 mag., so that the random error in the temperature determination is about 50K. However, the nominal accuracy of the color index is only realised when an homogeneous set of measurements is used. Moreover the systematic error of the Johnson's calibration could well reach also 50K. As an example, let us compare Johnson's and Popper's calibrations : for the color index *V-R* = 0.54, Johnson's and Popper's calibrations give respectively T_{eff} = 5660K and 5780K (the difference amounts to 120K). Other calibrations exist, such as the one by Barnes and Evans (1976), and some other ones, especially for giants (see Schmidtke et al. 1986 and references therein). Studies of the calibrations of the color indices for dwarfs, especially of the influence of metallicities, have been made by several authors (Peterson and Carney 1979, Carney 1983). Saxner and Hammerbäck (1985) provided a calibration from the application of the Infrared flux method , after Blackwell and Shallis (1977). The very interesting method of Blackwell et al. (the so-called "Infrared flux method") is supposed to provide for solar type stars an error of about 2 or 3% (or about 150K).

I cannot refrain from noting that when many calibrations are proposed, it means that a really good one does not yet exist, apart from the fact that it is normal that a particular calibration may be better fitted for a particular range of temperature. The determination of T_{eff} can be made by the adjustment of the profiles of the Balmer lines (Baschek 1959, Gehren 1981, Cayrel et al. 1985a, Cayrel and Bentolila 1988); such profiles are very sensitive to the temperature, so that the method is very promising (Fig. 5). Evaluating the errors of the measurements, made on high quality spectra obtained with modern detectors, the authors (Cayrel et al. 1985a) find for the fitting of Hα profiles, an accuracy of about 20K, which is a major

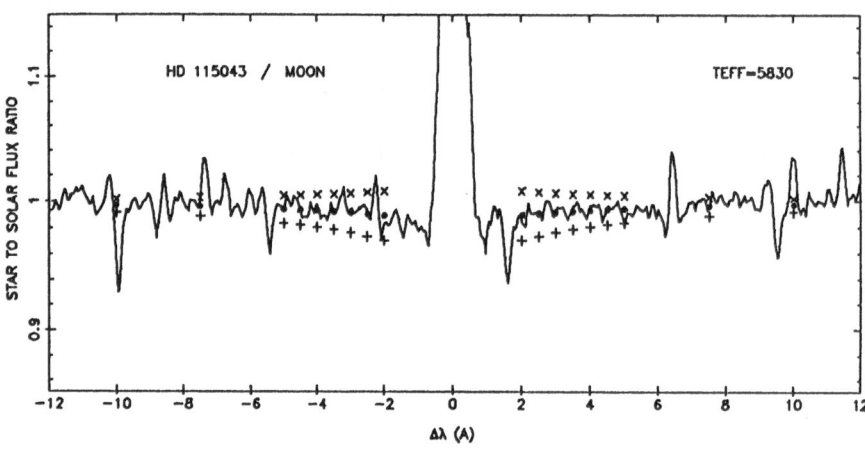

Fig. 5 . Ratio of stellar to moon spectra in the H_α region. Comparison with theoretical ratios for T_{eff} = 5380 K and Teff = 5830K $\pm$ 100K.The method provide a high precision in the determination of the effective temperature : see Cayrel (1988). Only the wings of the line are used, since the core is altered by the chromospheric activity. It remains to be checked that the wings themselves are not altered.

improvement. One problem is the compensation of the curvature of the continuum, which is difficult to achieve with precision. Another problem is, again, the limited spectral range obtained on the modern detectors; in the future, solutions may be found in the adoption of a lower resolution, and/or the installation of larger detectors (4096 pixels RETICON, mosaic of CCD etc...) combined with a spectrograph free of any kind of curvature of the continuum (ripple effect, vignetting etc...). In view of the considerable improvements realised by this method, such an effort would be extremely useful. A final problem is that, in solar observations, the wings of the Balmer lines could be changed through emission due to the chromospheric activity. This point has to be investigated before adopting the method once for all.

Let us note that both observations (colors and wings of Balmer lines) are relative to stellar layers of about the same depth, if the wings of the Balmer lines are not altered by the chromospheric activity.

With accurate line measurements, the excitation equilibrium for an element providing numerous observable lines (such as iron) can be well determined, providing (indirectly) a well defined value of T_{eff}. The lines are formed in the upper layers of the stellar atmosphere.

Table 1

Effective temperature (K)

colors	FeI	adopted
5840	5910	5880
5790	5660	5730
5280	5280	5280
5800	5920	5860
6120	6150	6140
6380	6420	6400
6320	6220	6270
5800	5870	5840
5720	5720	5720
6260	6070	6170
6060	5950	6010
5900	5920	5910
6050	5920	5990
6760	6660	6710
6710	6090	6130
5340	----	5340
5250	5360	5310
6240	6060	6150
6160	5910	6040
5310	5280	5300

Table published by Clegg et al. (1981)

Ideally it would be important to determine T_{eff} from the three methods for a few, well studied stars : from the agreement of the three values, an estimation of the adequacy of the model could be derived. However, in the literature, generally only two (at most) of these methods are used. For example let us consider the work of Clegg et al. (1981) : the agreement of the values of T_{eff} found from the colors and from the excitation equilibrium is about ± 100K, which is considered as satisfactory, since the differences are not systematic (Table 1).

A typical error of T_{eff}, when classical methods are used, amounts up to 200 K, corresponding to errors ranging from 0 to 0.1 dex in abundance.

3.4.3 Determinations of log g

a) In a purely spectroscopic determination, the gravity retained is the one which provides the ionization equilibrium, i. e. the equality of the abundance found from the neutral lines on one hand and the ionized lines on the other hand. For this determination, the iron lines (most of them are neutral lines) and the titanium lines (most of them are ionized lines) are mainly used because iron and titanium produce particularly numerous lines in both stages of ionization.

b) In some cases it is possible to derive the gravity by comparing the profiles of a faint and a strong lines of the same element, the lines originating from the same level. Schematizing the method, the shape of the wings of the strong line provides the gravity (see for example Edvardsson,1988). The gravity found by this method, if different from the gravity derived from the ionization equilibrium, is probably preferable.

c) When data about the luminosity and age of the star are available (for example if the star is member of a cluster), the mass and the radius of the star my be found from the theory of stellar evolution, and the gravity can be computed. In the case of the nearby star Procyon,which has a good parallax, Steffen (1985) finds a gravity which is not compatible with the ionization equilibrium.

Typical errors seem to be about $\Delta \log g$ = 0.2 dex, leading to an error of 0.0 to 0.1 dex in abundance.

4. A check of the abundance determinations

The abundances found in the Sun and in the meteorites are in a rather good agreement. Let us recall that the abundances determinations in the Sun suffer directly the uncertainties of the gf values.

5 .The abundance accuracy

As discussed hereabove, a classical temperature error in spectral analysis amounts easily to 200K. To such a temperature-error corresponds an abundance-error which is different for each line (according to the excitation potential) but may be evaluated in most cases to about 0.2 dex.

The Fe/H Catalog (Cayrel et al. 1985b) shows a rather large dispersion in the values of Fe/H. Of course, after discarding the less accurate analyses, the remaining scatter decreases, but still remains non negligible.

Similar evaluations are provided in the paper of Bikmaev (1988) : they are reproduced in Table 2.

For Procyon, the abundance error due to the inadequacy of the model could be as high as 0.17 dex (Steffen, 1985).

In general, in an analysis, the errors due to the imperfection of the model itself could amount to 0.1dex for the Sun (Blackwell and Shallis 1979).

ТАБЛИЦА 2

Эффективные температуры F-карликов, полученные прямыми и полупрямыми методами, К

BS 1543 π^3 Ori $V = 3.18$ F6 V	BS 2943 α CMi $V = 0.37$ F5 IV—V	BS 5447 σ Boo $V = 4.47$ F2 V	Ссылка
6470	6640	6750	[15]
6373	6600	6696	[15]
6600	6600	6935	[16]
6150	6278		[16]
	6500		[19]

Table published by Bikmaev (1987)

Gustafsson and Nissen (1978) estimate that the uncertainties entering into the synthetic spectra computations may sum up to more than 0.1dex.

Finally the total error can be estimated for each abundance determination by :

$\Delta A = (\partial A/\partial T)\, \Delta T + \partial A/\partial \log g\ \Delta \log g$ and the random error has to be added.

6. Relative abundances

The abundances derived from different lines of the different elements react differently to a change of the parameters. However in most cases the behavior is rather similar, so that relative abundances are less sensitive to errors in the determination of the temperature and of the gravity, and the obtained values are more reliable.

For illustration, let us show the example of the work by Tomkin et al.(1985). The authors find, by a comparison with a previous work of Wallerstein (1962) a non negligible difference for Fe/H (in accordance with our evaluation in Section 5), but generally negligible differences for relative abundances (Table 3).

Table 3

Abundance differences

Δ(Fe/H)   = -0.20 $\pm$ 0.06

Δ(Na/Fe) = +0.03 $\pm$ 0.10

Δ(Mg/Fe) = +0.08 $\pm$ 0.08

Δ(Si/Fe) = +0.08 $\pm$ 0.07

Δ(Ca/Fe) = +0.05 $\pm$ 0.08

Δ(Sc/Fe) = +0.06 $\pm$ 0.12

Table published by Tomkin et al. (1985).

Let us also find from this work an evaluation of the errors about relative abundances. The stars of about solar metallicity have nearly the same relative abundances of the metals (this was one of the conclusion of the work of Tomkin et al.). If we attribute to errors only the scatter around the mean, we get an upper limit of the errors. Depending of the element considered, the standard error (σ) amounts from 0.07 up to 0.13 dex. Obviously , some cosmic scatter exist in the relative abundances, and the errors are in fact smaller. The small amplitude of the errors is encouraging. The ratios which are especially independent of the models are of course most interesting. This is sometimes the case (but not always) for isotopic ratios. More generally, the case where a nearly complete compensation occur, should be especially studied, both from the theoretical and the observational point of view.

7. Molecular bands

Lambert (1978) estimates that, for the Sun, the abundance uncertainties due to the imperfection of the model atmosphere, is about 0.1 dex, an amplitude similar to the one estimated for the atomic lines.

However, here we have the additional difficulty that the dissociation energies are not always well known. Also the abundances derived from the molecular features are expected to be more sensitive to the gravity, and therefore to any gravity error. The table 7 of Clegg et al. (1981) shows, on an example, the obtained

8. UV and IR

The UV domain is a difficult one, as we have already shown for the Sun. The IR domain is promising. Some check of the temperature could even be found in this region. Quoting Gustafsson (1988) we note that since

1) the IR spectra of cool stars is relatively clean,

2) LTE is likely to prevail for the excitation of vibration-rotation levels of the electronic ground states of the molecules

3) the relative oscillator strengths are well known

it is tempting to use the strength of lines in different vibration-rotation bands of the diatomic molecules as temperature probes of the atmospheres

9. NLTE analyses

NLTE analyses should remove at least a part of the imperfections of the models. The problem is however not yet ready to be solved, since the cross sections for the numerous processes involved are generally lacking at present; moreover, the numerical difficulties should not be overlooked.

10. Peculiar stars

The normal stars are analysed with models which are built with the help of a firm comparison : the Sun. For peculiar stars (for example the carbon dwarf star analysed by Gass, Liebert and Wehrse, 1988), the situation is much more difficult. Therefore the errors could be significantly larger.

11. Conclusion

The accuracy of the abundances are of course especially interesting when compared to some variations observed : the problem is to find if these variations are significantly larger than the uncertainties. The answer is obviously important, since it has some consequences on the nucleosynthesis theory, the evolution of the Galaxy, etc...

Considerable work is in progress on the theoretical side. NLTE effects have been considered (for example Steenbock and Holweger 1984), the convection with inhomogeneities and an overionization effect is considered (Nordlund 1984), bringing improvements which could possibly reach up to a factor of two (but still very uncertain) in the solar iron abundance. No doubt that the careful comparison of different methods will help to find the right procedures for determining the abundances in a much safer way.

REFERENCES

Alexander J. B.:1986 *On the interpretation of Strömgren photometry* Month. Not. R. Astr. Soc. **220**, 473

Arpigny C., Magain P.: 1983 *The ratio Al/Mg in halo stars* Astron. Astrophys. **127**, L7

Barnes T. G.,Evans D. S.: 1976 *Stellar angular diameters and visual surface brightness* Month. Not. R. Astr. Soc.**174**, 489

Baschek B.: 1959 *Aufbau und chemische Zusammensetzung der atmosphäre des subdwarfs HD 140283*, Zeits. für Astrophys. **48**, 95

Beckers J. M., Bridges C. A., Gilliam L. B.:1976 *A high resolution spectral atlas of the Solar irradiance from 380 to 700 nm*, AFGL-TR760126, Hanscom, Mass.

Issled., Izv. Spets. Astrofiz. Obs., Tom.25, 3

Blackwell D. E., Shallis M. J.: 1977 *Stellar angular diameters from infrared photometry. Application to Arcturus and other stars; with effective temperatures* Month. Not. R. Astr. Soc. **180**, 177

Blackwell D. E., Booth A. J., Petford A. D., Leggett S. K., Mountain C. M., Selby M. J. : 1986 *The infrared flux method and its use for study of α Boo, μ Her and β Dra; relation to the Vega 1.2 - 5μm infrared excess* Month. Not. R. Astr. Soc. **221**, 427

Campbell B.: 1984, *Color anomalies and starspots in Hyades dwarfs* Astrophys. J. **283**, 209

Carney B. W.:1983 *A photometric search for halo binaries* Astron. J., **88**, 623

Cayrel de Strobel G., Bentolila C.:1988 in press

Cayrel R., Jugaku J.: 1963 Ann. Ap. **26**, 495

Cayrel R.:1988 *Data analysis* in IAU Sympos. 132(see Cayrel de Strobel G. and Spite M.) p. 345

Cayrel R., Cayrel de Strobel G., Campbell B. :1985a *Steps towards the abundance scale. I. The abundance of heavy elements in the Hyades cluster* Astron. Astrophys **146**, 249

Cayrel de Strobel G., Bentolila C., Hauck B., Duquennoy A.: 1985b *A Catalogue of Fe/H determinations* Astron. Astrophys Suppl. **59**, 145

Cayrel de Strobel G., Spite M., eds : 1988 *The impact of very high S/N spectroscopy on stellar physics*, IAU Sympos. **132**, Kluwer, Dordrecht

Clegg R. E. S., Lambert D. L., Tomkin J.:1981 *Carbon Nitrogen and Oxygen abundances in main sequence stars. II. 20 F and G stars* Astrophys. J. **250**, 262

Conti P. S., Deutsch A. J.: 1966 *Color anomalies and metal deficiencies in solar-type disk-population stars* Astrophys. J. **145**, 742

Delbouille L., Neven L., Roland G. :1973 *Photometric atlas of the Solar spectrum from λ 3000 to λ 10000*, IAP, Univ. Liège, Cointe-Ougrée

Dravins D.:1988 *Stellar granulation and photometric line asymmetries* in IAU Sympos. **132** (see Cayrel de Strobel G., Spite M.) p. 239

Edvardsson B.: 1988 Astron. Astrophys. *in press* and IAU Sympos. **132** (see Cayrel de Strobel G., Spite M.) p. 38

Gallouët L.: 1964 *Magnitudes stellaires du Soleil* Ann. Astrophys. **27**, 423

Gass H. Liebert J., Wehrse R.:1988 *Spectrum analysis of the extremely metal-poor Carbon dwarf star G 77-61*, Astron. Astrophys (to be published).

Gehren T.:1981 *The temperature scale of solar-type stars* Astron. Astrophys **100**, 97

Griffin R. F.: 1968 *A photometric atlas of the spectrum of Arcturus*, Cambridge Phil. Soc., Cambridge, UK

Griffin R., Griffin R.: 1979 *Photometric atlas of the spectrum of Procyon*, Institute of Astronomy, Cambridge, UK

Gustafsson B.:1988 *Physical input for the determination of stellar abundances* in IAU Sympos. **132** (see Cayrel de Strobel and Spite, 1988)

Gustafsson B., Bell R. A., Eriksson K., Nordlund, A.: 1975 *A grid of Model atmospheres for metal-deficient giant stars. I.* Astron. Astrophys. **42**, 407

Gustafsson B., Nissen P. E.: 1978 *An investigation of microturbulence and metal abundances in F dwarfs* Astr. papers dedicated to Bengt Strömgren, A. Reiz and T. Andersen, eds., Copenhagen Univ. Observatory, p. 43

Holweger H., Müller E. A.: 1974 *The photospheric Barium spectrum : solar abundance and collision broadening of Ba II lines by Hydrogen* Solar Physics **39**, 19

Johnson H. L. : 1966 *Astronomical measurements in the Infrared*, Ann. Rev. Astron. Astrophys **4**, 193

Kron G. E.: 1963 *The color of the Sun* Pub. Astr. Soc. Pacific **75**, 288

Kurucz R. L.:1979 *Model atmospheres for G F A B and O stars* Astrophys. J. Suppl.**40**, 1

Kurucz R. L., Avrett E. H.:1981, *Solar spectrum synthesis I.- A sample atlas from 224 t o 300 nm*, Smithsonian Institution, Cambridge, Mass.

Kurucz R. L., Furenlid I., Brault J., Testerman L.: 1984, Solar flux atlas from 296 to 1111111 1300 nm, Nat. Sol. Obs., Sunspot, NM

LaBonte B. J., Rose J.A.: 1985 Publ. Astr. Soc. Pacific **97**,790

Lambert D. L.: 1978 *The abundances of the elements in the solar photosphere. VIII. Revised abundances of CNO* Month. Not. R. Astr. Soc. **182**, 249

Magain P.: 1985 *Spectroscopic analysis of extreme metal-poor dwarfs. II. Improved model atmospheres and detailed abundances* Astron. Astrophys. **146**, 95

Nissen P. E.:1988 *The fourth parameter problem in uvby-β photometry of open clusters* Astron. Astrophys. **199**, 146

Nordlund A.:1984 in: *Small scale dynamic processes in quiet stellar atmospheres*, S. L. Keil, ed., Sacramento Peak, p. 30

Peterson R. C., Carney B. W.:1979, Astrophys J. **231**, 762

Popper D. M.: 1980 *Stellar masses* Ann. Rev. Astron. Astrophys. **18**, 115

Saxner M., Hammarbäck G.:1985 *An empirical temperature calibration for F dwarfs*, **151**, 372

Schmidtke P. C., Africano J. L., Jacoby G. H., Joyce R. R., Ridgway S. T.:1986 *Angular diameters. by the lunar occultation technique VII*. Astron. J. **91**, 961

Spite F., François P., Spite M.: 1985 *Spectrophotometry of Globular Cluster stars with the CASPEC system; a comparison with results from other spectrographs* The Messenger N° 42, ESO, Garching

Steenbock W., Holweger H. : 1984 *Statistical equilibrium of Lithium in cool stars of different metallicity* **130**, 319

Steffen M.:1985 *A model atmosphere analysis of the F5 IV-V subgiant Procyon* Suppl. Ser. **59**, 403

Tomkin J., Lambert,D. L., Balachandran S.: 1985 *Light element abundances in 20 F and G stars* Ap. J. **290**, 289

Wallerstein G.: 1962 *Abundances in G dwarfs. VI. A survey of field stars* Astrophys. J. Suppl. Ser. **6**, 407

APPENDIX : Evaluation of the parameters of the model

Since the optimum fitting of observed and computed spectra does not provide a very sharp determination of the model parameters, the usual procedure is to try to determine separately (as far as possible) each parameter.

Let us recall the classical steps for an (LTE) abundance determination :

1) Determination of the effective temperature

The effective temperature can be deduced from the colors, the profile of the Balmer lines, or from the excitation temperature.

About the colors, it is well known that the models do not predict completely correct colors, athough the general trend is correct. The relation between colors and temperature is therefore indirect but it is generally quite good for dwarfs and giants. Supergiants are generally far away in the Galaxy, and are reddened; the reddening correction is often uncertain; so that the intrinsic colors of the supergiants are not very well known, and the relation between color and temperature is not very accurate.

Balmer line profiles can provide the effective temperature of the star. The experimental difficulty is the determination of the continuum level. By using ratios of several depths in the profile, the error in the location of the continuum can be largely attenuated. The remaining difficulty is the possible alteration of the profiles by chromospheric emission.

The effective temperature of the star can be deduced from the excitation equilibrium, i. e. the condition that the individual values of the abundance of an element derived from individual lines, should be independent of the excitation potential of the lower level of the transition corresponding to the measured line.

Each of this method is largely independent of the other ones. In particular, the method based on the use of excitation potentials is relative to measurements of lines formed in rather superficial layers of the star (corresponding on a continuum depth log $\tau = 0.1$). The other methods rely on the measurement of fluxes originating essentially in deeper layers of the star (continuum depth log $\tau = 1$).

Obviously, a good check of the adequacy of the models is that the three methods provide temperature values in agreement.

This is generally the case for cool dwarfs.

It has to be noted that, obviously, the *observables* used in the three methods listed hereabove are not dependent of the temperature **only** : they depend slightly of the gravity and of the metallicity of the star; so that an iteration process has to be used; this is easy to perform with modern computers.

2) Determination of the gravity

In a purely spectroscopic determination, the gravity retained is the one which provides the ionization equilibrium, i. e. the equality of the abundance found from the neutral lines on one hand and the ionized lines on the other hand. For this determination, the iron lines (most of them are neutral lines) and the titanium lines (most of them are ionized lines) are mainly used because iron and titanium produce particularly numerous lines in both stages of ionization.

In some cases it is possible to derive the gravity by comparing the profiles of a faint and a strong lines of the same element, the lines originating from the same level. Schematizing the method, the shape of the wings of the strong line provides the gravity (see for example Edvardsson,1988).

When data about the luminosity and age of the star are available (for example if the star is member of a cluster), the mass and the radius of the star my be found from the theory of stellar evolution, and the gravity can be computed.

3) Determination of the abundances of the elements

For the faint lines, the equality of the equivalent width (or of the profile) of a computed line and of the observed line determines the abundance of the element, assuming that the value of the atomic constant *gf* is known. The uncertainty of these constants are one of the cause of the limitation of the accuracy of the abundance determination. In the past some of the laboratory values have been plagued by systematic errors. But progresses have been made (for example Blackwell and collaborators, Huber, etc...).

A differential analysis relative to the Sun is in principle independent of the laboratory systematic errors. But the differential method turns out to be a direct comparison of the stellar line to the solar line, and if the lines are not in the same state of saturation, a correction has to be made, which will depend on both the solar and the stellar models.

We call here "faint lines" the lines which are on the linear part of the curve of growth. As soon as a line is strong enough for being affected by the desaturation effect, this effect has to be taken into account, and it is necessary to proceed to another step :

4) Determination of the microturbulence parameter

Owing to the velocity field of the stellar atmospheres, the lines are de-saturated by Doppler displacements. The velocity field is usually supposed to be the consequence of the convective movements of the stellar matter below the atmospheres. The parameter describing the desaturation is classically given in terms of velocity and is named ξ or v_t This parameter is determined by the condition that the abundance of an element should be independent of the strength of the line used for the abundance determination.

It could be thought that it is enough to use only faint lines, which are affected neither by damping nor by desaturation, and this is true if an element is represented in the spectrum by numerous faint lines. But these faint lines are not always numerous for a given element and are even completely absent for some elements. It is true that the lines affected by damping and/or saturation do not bring basically new information on the abundances, when faint lines are available. A good determination of ξ brings in agreement the results obtained from both faint and strong lines, so that the strong lines can be used as well as faint lines for reliable abundance determination.

5) The damping

As soon as the wings of a line play a significant role in the strength of the line, the pressure damping has to be taken into account. Only approximate formulae for the damping effect are available up to now.

PROBLEMS ASSOCIATED WITH COOL DWARF STARS

M.S. Bessell[1] and M. Scholz[2]

[1] Mount Stromlo and Siding Spring Observatory
Canberra, Australia
[2] Institut für Theoretische Astrophysik
der Universität Heidelberg, Heidelberg, F.R.G.

Abstract. Models for cool stars differ from those for hotter stars by having complicated state equations and opacities dominated by lines from diatomic and polyatomic molecules. In addition, in cool dwarfs most of the energy transport is by convection, except in the uppermost layers, and methods of handling convection, particularly far away from the adiabatic limit is inadequate. The $T(\tau)$ structure of cool dwarfs varies with temperature, gravity and metallicity; therefore scaled solar models cannot be used in analyses. The effective temperature scale of cool dwarfs is also not well known, there being few stars with fundamental temperature measurements.

I. HISTORICAL OVERVIEW

Early attempts at constructing model atmospheres for cool stars were undertaken by Auman (1967,1969), Tsuji (1968) and Mould (1976, 1978) (M dwarfs), and by Tsuji (1966), Alexander et al (1972), and Johnson (1974) for M giant stars.

Various methods of handling the complex line absorption opacities have been considered, mean opacities (eg Golden 1969; Tsuji 1966; Zeidler et al, 1982), Elsasser Band model (Tsuji 1968,1978), opacity distribution functions (ODF)(eg Querci et al 1971,1974; Gustaffson et al, 1975; Saxner et al, 1984) and opacity sampling (OS) method (eg Peytremann,1974; Sneden et al,1976).. More recent plane-parallel geometry models for M giant stars have been published by Tsuji (1981), and Johnson et al (1980,OS).

Mould (1976) published the most recent grid of M dwarf models, and although the published models cooler than 3000K have H_2O bands much stronger than observations (Persson et al 1977), his unpublished models with H_2O arbitrarily diminished to 1/3 are in better agreement with observations. Bell and Gustafsson (1980) have extended their grid of giant models (Gustafsson et al 1975) to cover the gravities of the dwarf stars, and although the grid is not published the models are available. Kurucz and Buser (1988) have also computed fluxes for these dwarf models and incorporated Kurucz's more extensive line lists. These models are relevant for G and K dwarfs, but not M dwarfs, as they contain no TiO or H_2O opacities. They supplement the hotter grid of dwarf models (Kurucz 1979a,b).

I. MODELLING PROBLEMS CONCERNING ALL COOL PHOTOSPHERES

1. Obtaining the opacities.

(i) Cool stars have complicated state equations with large number of composed particles (including dust at very low temperatures). (ii) There are possibly missing particles (in particular at very low temperatures) in the state equation which may lead to spurious particle pressures of the absorbing particles. (iii) The physical input data of the state equation may be poorly known (cf. revisions of TiO dissociation energy in the past) which also will lead to spurious partial pressures of the absorbing particles. (iv) The absorption is dominated by very large numbers of lines, in particular molecular band lines (plus dust at very low temperatures). (v) Missing absorbers (in particular at very low temperatures) may lead to grossly incorrect opacities, and the oscillator strengths of important bands may be unknown or poorly known (cf. several TiO bands).

2. Handling the opacities.

(i) The continuum opacity (H^-, H_2^-) is fading away towards low temperatures and opacities become line-dominated. (ii) Treatment of molecular lines by the mean-opacity technique reduces the required computer capacity (storage, time) but may be seriously inadequate, in particular for saturated lines. (iii) Treatment of lines by the ODF or the OS techniques requires large computer capacity. OS is more flexible and enables the treatment of macroscopic velocity fields. (iv) Complicated molecules (e.g. H_2O, HCN and C_3) with complicated line patterns are treatable with theoretical models of the molecule (e.g. Jørgensen et al 1985) or by statistical derivation of line position for synthetic spectra (e.g. Alexander et al 1988) but will require even more computing capacity and the use of super-computers.

3. Modelling and analyzing the photosphere.

(i) Strong wavelength dependence of opacities may lead to difficulties in finding the stratification which fulfills the energy equation, in particular via temperature correction methods which use weighted mean opacities (e.g. Lucy). (ii) Very accurate modelling is required if an important absorber which has great influence upon the stratification reacts sensitively to stratification changes (e.g. H_2O). (iii) Gravity sensitive features are hard to find at very low temperatures. (iv) Since all elements are coupled with each other via the composed particles in the state equation, a consistent set of element abundances has to be determined instead of carrying out an element by element analysis. (v) Since absorption is dominated by lines, the abundance input for model construction and the abundances deduced from analysis have to be checked carefully for consistency. Not even the continuous absorption is a safe pre-given quantity because the free e^- forming H^- and H_2^- depend on the abundances of the alkali-earth and the alkali metals (which are less easily accessible than the Fe-type metals supplying the free e^- in hotter stars). (vi) An element which is not observable may still be important concerning the state equation and the electron donation.

II. MODELLING PROBLEMS CONCERNING KM DWARF PHOTOSPHERES

1. Modelling a photosphere dominated by convective energy transport.

Treatment of convection is usually by a mixing-length method. This is scarcely adequate and modifications have been attempted (e.g. Lester et al 1982). There is also no adequate treatment of convection which is far away from the adiabatic limit. There are also difficulties in finding T(r) in the uppermost *radiative* layers $r > r_0$. This is extremely important, because an exact boundary temperature $T(r_0)$ is required so that the correct T(r) below r_0 can be found as convection determines essentially the dT(r)/dr gradient.

2. Temperature stratifications.

For F and G stars a scaled solar $T(\tau)$ scale is adequate for most abundances and gravities, but for late G and K dwarfs, particularly for low metallicity compositions, scaled solar $T(\tau)$ relations cannot be used. In G and K dwarfs between 6000K and 3800K the $T(\tau)$ gradient undergoes great variations. In Figure 1 are shown some unblanketed dwarf model stratifications from Bessell et al (1979). The temperature gradient decreases as (i) the temperature is decreased at a given gravity and abundance, (ii) the gravity is increased at a given temperature and abundance, (iii) the abundance is decreased at a given temperature and gravity. This is caused by the increase in the importance of convective energy transport. For temperatures hotter than the sun, the atmospheres are radiative, and for M dwarfs, 3500K and cooler, all dwarf atmospheres are nearly fully convective.

3. Derivation of gravity.

Precise gravities are difficult to determine for GK dwarf stars. In hotter stars the Balmer Jump can be used for gravity determinations but this becomes insensitive the cooler the temperature becomes. The damping wings of strong lines can be used, but the electron pressure is sensitive to metallicity as well as to gas pressure or gravity. Accurate measures can be made for bright, normal K dwarfs but are difficult for faint metal-poor stars.

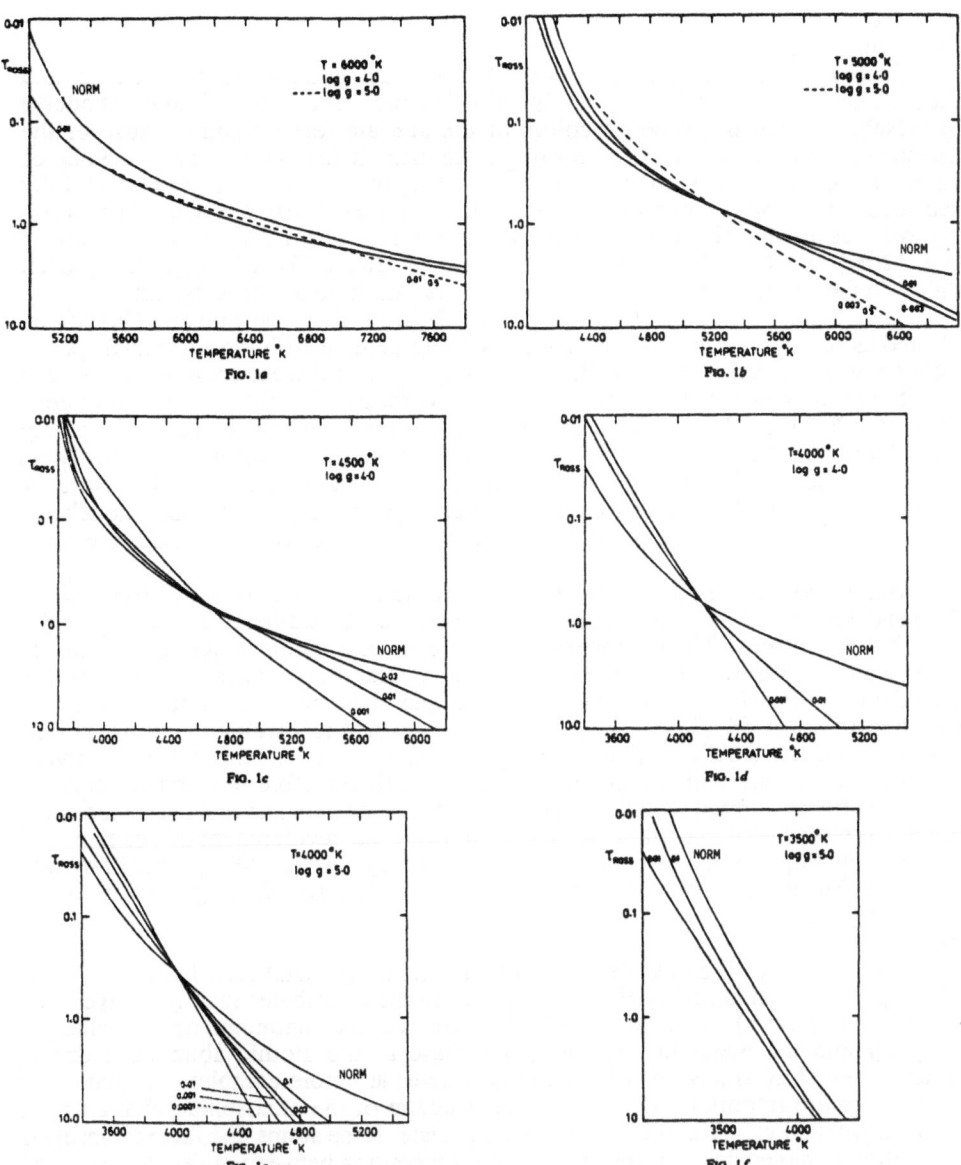

Fig. 1.—The T-τ relationships for some models illustrating the changing structure of the atmospheres as the temperature is reduced, the gravity increased and the metal abundance decreased.

4. Inhomogeneous atmospheres.

There are significant complications in cool stars due to possible spatial inhomogeneities, such as photospheric granulation (Dravins 1987), star spots and plages, and due to chromospheres. These effects are presumably most severe in solar-type G and early K stars and less severe in the fully convective M dwarfs, however many of the late-M dwarfs are flare stars and some show evidence for a filling-in of strong TiO bands in the visible. Most very late-type M dwarfs are variable in light and to a lesser extent in color, at least in the VRI passbands.

III. OBSERVATIONAL ASPECTS OF COOL DWARFS
1. Temperature determinations.

Because of their small radii, few dwarf stars have had their radii measured, and thus their effective temperatures derived directly. The radii of the sun is known of course very precisely, but the photometric colors of the sun are less certain because of the instrumental difficulties involved in comparing a -26 magnitude day-time star of 0.5 degree angular extent, to night-time point sources of 0 to 5 magnitudes. However, use of carefully selected solar-analog stars can minimize these uncertainties. There are two other dwarfs which have measured radii, these stars are eclipsing binaries; YY Gem, a M0.5 dwarf (3780 ± 120K) and CM Dra, a high velocity M4.5 dwarf (3150±50K). The greater uncertainty in the temperature for YY Gem results from its more uncertain parallax.

The VRI colors computed by Kurucz et al (1988) from the Bell and Gustafsson (1980) models for the G and K dwarfs are in good agreement with the solar analogs and extrapolate toward the values for YY Gem. This suggests that these models and colors can be used for interpretation of G and K dwarfs. For M dwarfs, Mould's fluxes are in good agreement for YY Gem but deviate for CM Dra; however, the corrected H_2O opacity models (Mould 1978) are in better agreement with the observations of CM Dra. These models can be used for interpretation of the early M dwarfs. It is difficult to assess the dwarf models of Tsuji (1968) as few details have been published, but his giant models are in excellent agreement with the fundamental temperatures of M giants (Ridgway et al (1980).

Black-body fluxes have been used to derive temperatures for M dwarfs. Reid et al (1984) and Veeder (1974) fitted essentially the integrated broad-band JHKL magnitudes with black-body curves. Their temperatures were in reasonable agreement with the fundamental temperatures for YY Gem and CM Dra, but because of the grossly non-blackbody fluxes of M dwarfs the agreement may have been fortuitous. Wing et al (1980) fitted the near-IR (0.8μ - 1.1μ) flux of M dwarfs to blackbody curves. Their derived temperatures were also in reasonable agreement with the fundamental temperatures.

For dwarf stars with temperatures above 3100K therefore the temperatures are known at least for the normal metallicity stars, but for cooler stars and metal deficient stars the temperatures are very uncertain and we must await the development of better model atmospheres and hope for the discovery of more eclipsing binaries. Models for M dwarfs are under construction by Tsuji (Tokyo), Allard and Wehrse (Heidelberg) and Ruan and Bessell (ANU).
2. Spectral peculiarities.

The spectra of F and G dwarfs contain essentially metal (and hydrogen) lines only. The opacity is dominated by H^- and the electrons are contributed mainly by hydrogen (and Fe in late G dwarfs). As the metallicity is decreased, the continuum opacity changes only slightly and the metal line strengths decrease as the atomic abundance drops. However, in late G, K and M dwarfs, the atmospheres are more complex. Formation of molecules becomes important resulting in the structure of the atmosphere changing (as discussed above) and via a complicated equation of state some atoms distribute themselves between different molecules and free atoms. The competition between molecules for some atoms, such as oxygen and carbon, can result in the CH molecular species increasing in abundance as the metallicity decreases. The number densities of molecules involving two metals, such as TiO, decrease rapidly as metallicity falls, while hydrides, such as MgH and CaH show much smaller decrease in density (Cottrell 1978; Bessell 1982). Bessell et al (1989) have obtained spectra of many high velocity cool dwarfs and some examples (Bessell 1982) are shown in Fig. 2. These illustrate the weakness of the TiO bands and the great strength of the MgH, CaH bands and Na and K lines. These stars are thought to have metallicities near 1/100 solar. The spectra are arranged with the hotter stars at the top.

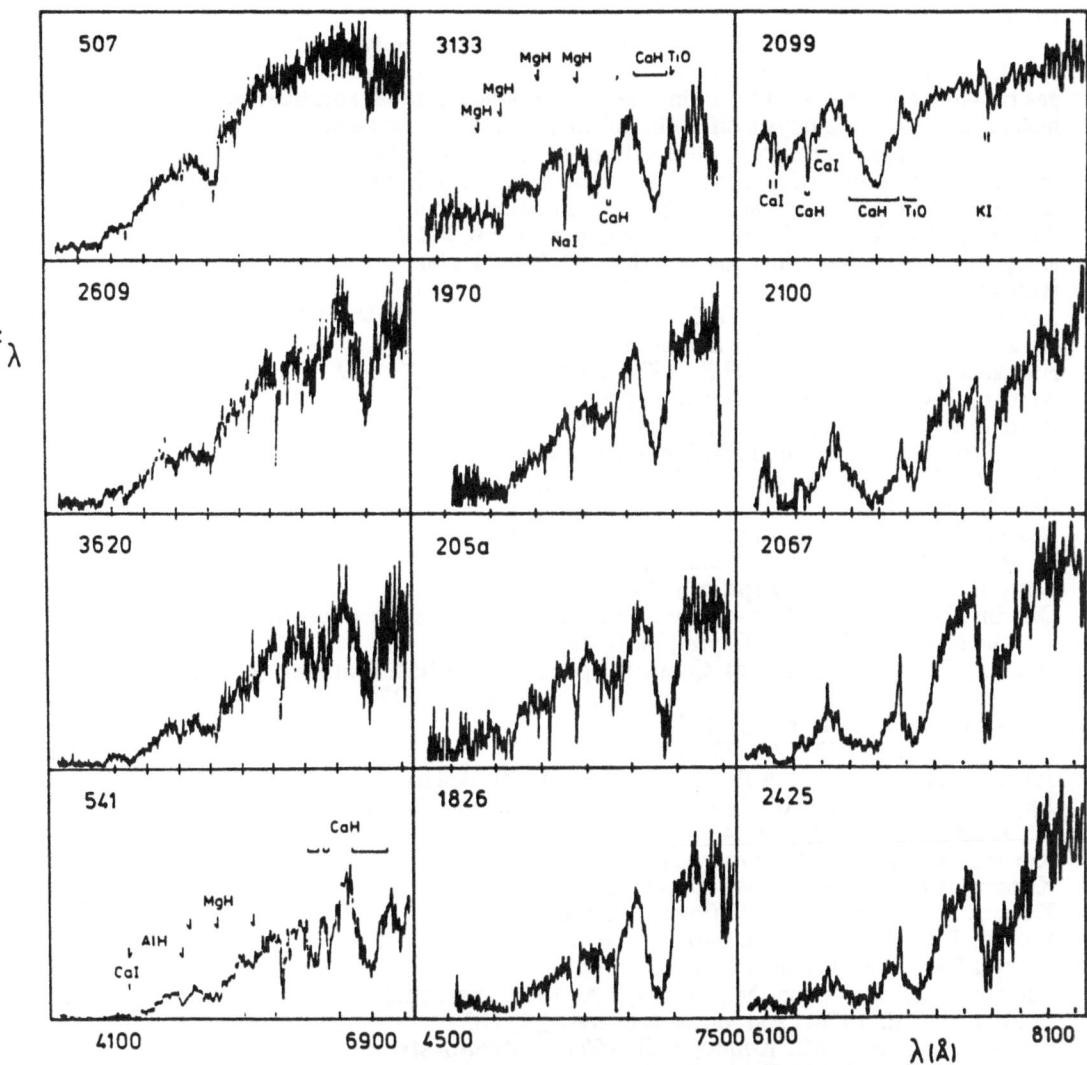

F_λ

Fig. 2.— The spectra of some red LHS stars believed to be cool metal-deficient dwarfs.

The increased association of H_2O molecules in metal-poor M dwarfs also results in strengthening of the IR H_2O bands and consequent change in the JHKL magnitudes. In addition, the H^- (and H_2^-)continuum opacity, which is proportional to the electron number density, also decreases as the metallicity decreases because the electrons come from elements like Fe, Mg (GK) and Na (M). This decrease in the continuum opacity results in spectral lines not weakening in KM dwarfs as much as they do in FG dwarfs as the metallicity drops. The gas pressure also increases as the metallicity falls increasing the collisional broadening of lines and producing a corresponding increase in the strength of strong metal lines, such as CaI 4226Å, NaD 5893Å, KI 7665Å and particular FeI lines. The net effect of all these interactions is that the spectra of KM dwarfs undergo dramatic, and at first glance, unexpected changes as the metallicity and temperature decreases. Even greater changes occur in higher gravity white dwarfs (Wehrse 1977; Bessell et al 1980).

IV. SUMMARY

The atmospheres of cool dwarfs change considerably more than those for hotter dwarfs as the temperature, gravity and metallicity change. Convection, molecule formation, gas pressure changes, all interact in ways that produce unusual spectroscopic changes and make spectroscopic analysis difficult; differential analyses are also much less certain than for hotter stars. Spectroscopic observations in the visible and IR are much more difficult than for most other stars because of their intrinsic faintness. Modelling the stars is complicated by convection and the great contribution of polyatomic H_2O to the opacity. All these effects make cool dwarfs a challenging problem, but if you need accurate abundance analyses at the present time, concentrate on B,A and F stars.

References

Alexander, D.R., Augason, G.C., Brown, J.A., Johnson, H.R. 1988,
    I.A.U. Colloquium 106 Bloomington, Indiana
Alexander, D.R., Johnson, H.R. 1972, **Astrophys.J. 176,** 629
Auman, J.R. 1967, **Astrophys.J.Suppl. 14,** 171
Auman, J.R. 1969, **Astrophys.J. 157,** 799
Bell, R.A., Gustafsson, B. 1980 (Private Communication)
Bessell, M.S. 1982, **Proc.Astr.Soc.Aust. 4,** 417
Bessell, M.S., Liebert, J. 1989 (In Preparation)
Bessell, M.S., Wickramasinghe, D.T., Cottrell, P.L. 1980 IAU Colloquium 53
    (eds: H.M.Van Horn, V.Weidemann) p179
Cottrell, P.L. 1978, **Astrophys.J. 223,** 544
Dravins, D. 1987 **IAU Symposium 132** (Kluwer Academic Publishers,
    The Netherlands ed. G. Cayrel, M. Spite) p239
Golden, S.A. 1969, **Journal Quantit.Spectrosc.Radiat.Transfer 9,** 1067
Gustafsson, B., Bell, R.A., Eriksson, K., Nordlund, A. 1975,
    Astron.Astrophys. 42, 407
Johnson, H.R. 1974 NCAR-TN/STR-95
Johnson, H.R., Bernat, A.P., Krupp, B.M. 1980, **Astrophys.J.Suppl. 42,** 581
Jørgensen, U.G., Almhof, J., Gustafsson, B., Larsson, M.,Siegbahn, P. 1985,
    J.Chem.Phys. 83, 3034
Kurucz, R.L.1979a, **Astrophys.J.Suppl. 40,** 1
Kurucz, R.L. 1979b, **Dudley Obs.Report** no **14,** 271
Kurucz, R.L., Buser, R. 1988 (Private Communication)
Mould, J.R. 1976, **Astron.Astrophys. 48,** 443.
Mould, J.R. 1978, (Private Communication)
Persson, S.E., Aaronson, M., Frogel, J.A. 1977, **Astron.J. 82,** 729.
Petreymann, E. 1974, **Astron.Astrophys. 33,** 203
Querci, F., Querci, M., Kunde, V.G. 1971, **Astron.Astrophys. 15,** 256
Querci, F., Querci, M., Tsuji, T. 1974, **Astron.Astrophys. 31,** 265
Reid, N., Gilmore, G. 1984, **Mon.Not.Roy.Astr.Soc. 206,** 19
Ridgway, S.T., Joyce, R.R., White, N.M., Wing, R.F. 1980, **Astrophys.J. 235,** 126
Saxner, M., Gustafsson, B. 1984, **Astron.Astrophys. 140,** 334
Sneden, C., Johnson, H.R., Krupp, B.M. 1976, **Astrophys.J. 204,** 281
Steiman-Cameron, T.Y., Johnson, H.R. 1986, **Astrophys.J. 301,** 868
Tsuji, T. 1966, **Pub.Astr.Soc.Japan 18,** 127
Tsuji, T. 1968, in **Low Luminosity Stars,** ed. S.Kumar, Gordon and Breach, NY, p457
Tsuji, T. 1978, **Astron.Astrophys. 62,** 29
Tsuji, T. 1981, **J. Astrophys.Astron. 2,** 95
Veeder,G.J. 1974, **Astron.J. 79,** 1056
Wehrse, R. 1977, **Mem.Soc.Astr.Italia. 48,** 13
Wing, R.F., Dean, C.A.1980, (Private Communication)
Zeidler-K.T., E.-M., Koester, D. 1982, **Astron.Astrophys. 113,** 173

ACCURACY OF ABUNDANCES FROM STARS IN NEAR-BY GALAXIES

B. Baschek
Institut für Theoretische Astrophysik
Im Neuenheimer Feld 561
D 6900 Heidelberg

ABSTRACT

Abundance determinations in near-by galaxies are discussed on the basis of analyses of *individual stars*. Accurate abundances for a large number of elements require high-resolution spectra with $R = \frac{\lambda}{\Delta\lambda} \geq 10^4$. Reasonable results can still be obtained with $R \simeq 10^3$. At these resolutions, confusion by other stars in the aperture is not a serious problem.

Compared to supergiants, near-main sequence B stars have the advantage that their analysis is not severely affected by NLTE effects and that their surface composition is not contaminated by processed matter. At present, these stars can be reached with $R \geq 10^4$ in the Magellanic Clouds. The very large telescopes of the 1990's will make them accessible in M31 with high resolution. As an illustration, spectra of two near-main sequence B stars in the LMC and SMC with $m_V \simeq 14$ mag , obtained with CASPEC at the 3.6 m telescope of ESO are presented, and preliminary abundance results are reported.

The dependence of the limiting magnitude at a given spectral resolution from telescope diameter, S/N, and exposure time is discussed, as well as the loss of accuracy and of information about chemical elements as a consequence of reducing the spectral resolution in order to reach more distant galaxies.

1. INTRODUCTION

The preceding contributions dealt with the accuracy of abundance determination which can at present be achieved for various types of stars essentially all of which belong to our own Galaxy. In view of the development of highly sensitive detectors and of very large telescopes which will be available in about a decade, it is also of interest to discuss how far high-resolution stellar spectra and detailed information on element abundances can be obtained from *other galaxies*. In this contribution, the discussion of abundances and their accuracy will be restricted to the results of spectroscopy of *individual stars*. Neither the wealth of data on abundances in HII regions nor the problems of population synthesis and the abundances derived by this technique in unresolved or only partly resolved stellar systems will be covered here (cf. e.g. Pagel and Edmunds, 1981).

Obviously, for "individual" stellar spectroscopy in galaxies one has to aim to observe stars as bright as possible in order to obtain high resolution spectra with a large signal-to-noise ratio. On the other hand, it is the most luminous stars such as supergiants and supernovae which at present set problems with reliable abundance determinations, e.g. due to uncertainties in the structure of extended or expanding atmospheres and in the calculation of occupation numbers in pronounced non-LTE situations.

A further important aim in the study of various types of galaxies is the determination of the chemical composition of unevolved stars which still exhibit their original element mixture. This can then be compared to the composition of the interstellar gas as determined e.g. from the emission spectra of HII regions. However, OB main sequence stars, being the brightest unevolved stars, are several magnitudes fainter than supergiants so that their analysis in other galaxies is restricted to the nearest systems even if the largest telescopes are employed.

We begin with an illustration of the present state of the art in the determination of extragalactic abundances from stars on the basis of recent results of our Heidelberg group about near-main sequence B stars in the Magellanic Clouds and luminous blue variables (LBV) in M31 and M33. After an outline of the limits to resolve individual stars in galaxies the future possibilities for abundance determinations opened by very large telescopes are discussed. Finally, we deal with the loss of accuracy and information on abundances due to decreasing spectral resolution which is unavoidable if more distant galaxies are to be observed.

2. EXTRAGALACTIC ABUNDANCES FROM INDIVIDUAL STARS - THE PRESENT STATE OF THE ART

As an example, first results of our Heidelberg collaboration are reported where we succeeded in securing high-resolution spectra of two *near-main sequence* B stars in the Magellanic Clouds with a good signal-to-noise ratio and in determining abundances for a number of elements. These observations represent the present limitations, typical for a telescope of the 3 to 4 m class with an efficient high-resolution spectrograph. In the following, these results obtained for the Magellanic Clouds will serve as a reference for scaling the spectroscopic possibilities in more distant galaxies.

Before giving some details, it seems worth-while to point out the advantages of the analyses of near-main sequence B stars as compared to the considerably brighter supergiants. O and B stars at or near the main sequence are the brightest unevolved objects. Their surface composition is not contaminated by nuclear processing and should reflect the composition of the interstellar gas out of which they formed. Hence a comparison of their abundances with those of HII regions is of prime interest. An additional advantage of choosing B stars instead of supergiants is that their abundance analysis is hardly affected by non-LTE effects, particularly if the analysis is performed *differentially* to a suitable standard star.

The observations by the Heidelberg collaboration (A. Reitermann, O. Stahl, B. Wolf, B. Baschek, M. Scholz) have been carried out with the echelle spectrograph CASPEC at the 3.6 m telescope of ESO in Chile. The CCD spectra cover the wavelength range 3900 to 4940 Å with a resolution $R = \frac{\lambda}{\Delta\lambda} \simeq 2\,10^4$, corresponding to $\Delta\lambda \simeq 0.25$ Å at 5000 Å. Since the brightness of the B stars in the Magellanic Clouds is only $\simeq 14$ mag, at least two spectra of 1.5 to 2 hours exposure each had to be superimposed in order to achieve a S/N ratio better than 50. The minimum equivalent width which can still be measured with fair accuracy is $W_{\lambda,min} \simeq 15$ mÅ.

The two (sharp-lined) B stars chosen for the analysis are both members of blue globular clusters, NGC 1818 in the LMC, and NGC 330 in the SMC. Their visual brightness V and

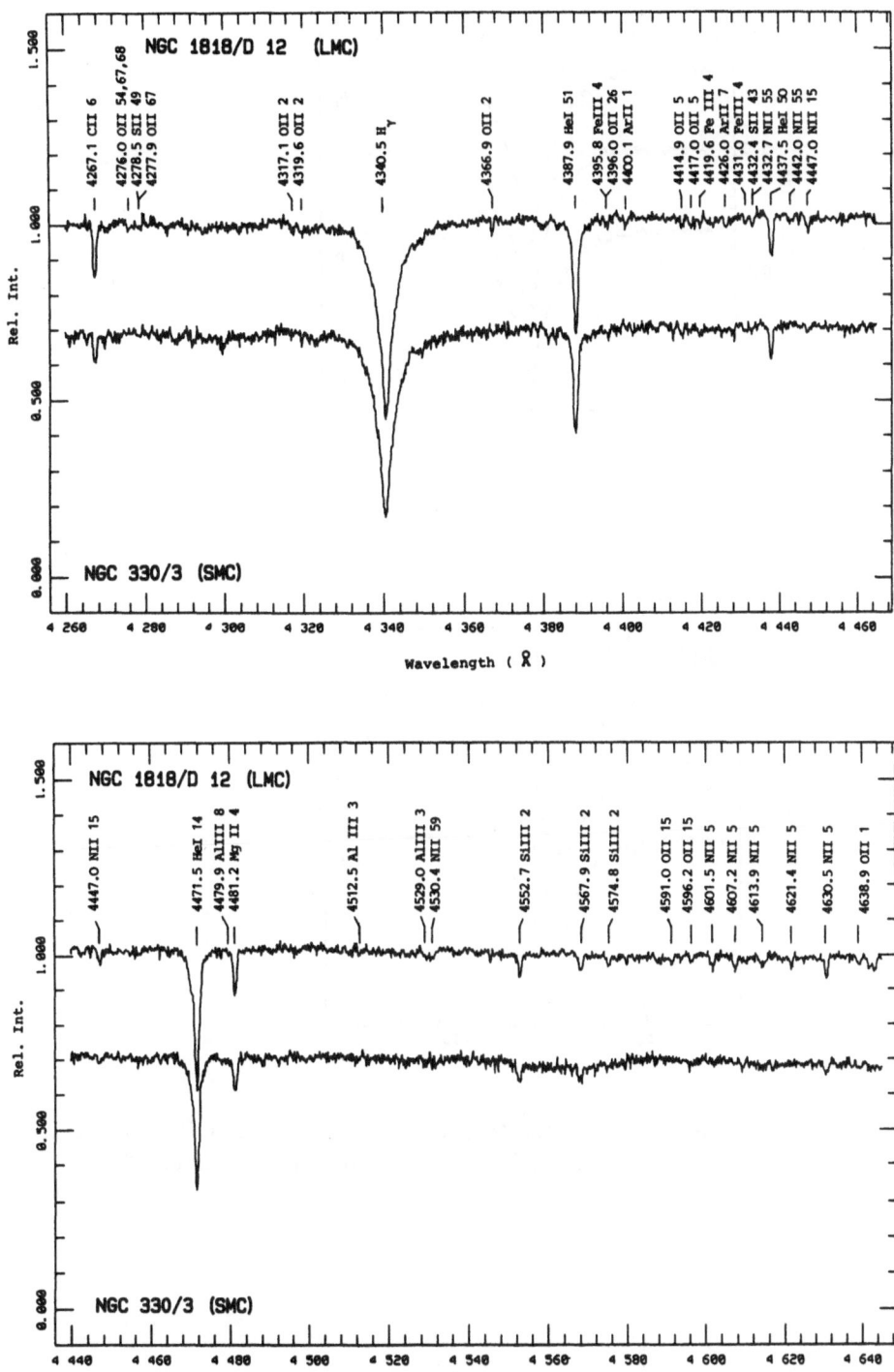

Fig. 1. The B stars NGC 1818/D12 and NGC 330/3 in the Magellanic Clouds. Sections of the spectrum with identifications (Reitermann, 1989)

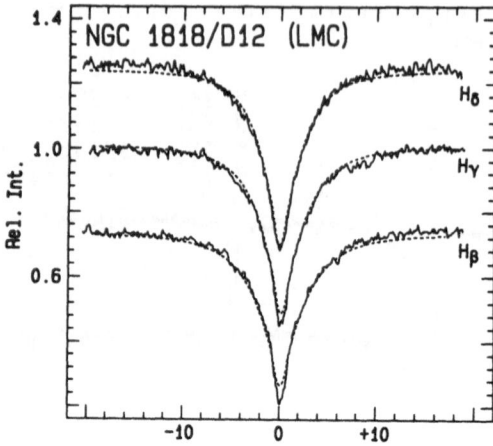

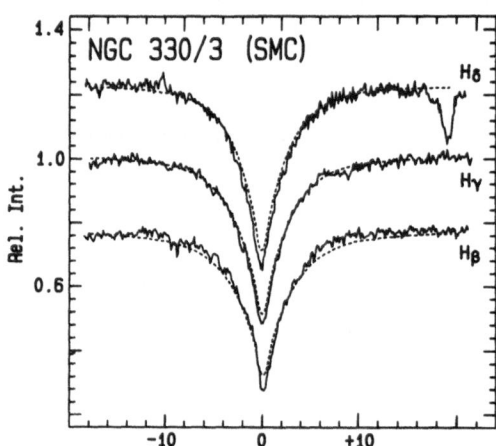

Fig. 2. Balmer Lines H_β, H_γ and H_δ in the Magellanic Cloud B stars NGC 1818/D12 and NGC 330/3. Observed and calculated (dashed lines) profiles (Reitermann, 1989)

derived model parameters, effective temperature T_{eff} and surface gravity g_{eff}, are:

star	NGC 1818/D12	NGC 330/3
galaxy	LMC	SMC
V[mag]	13.65	14.22
T_{eff}[K]	17900	21000
$\log g \, [\mathrm{cm\,s^{-2}}]$	2.85	3.25

The model parameters have been derived from the equivalent widths of the Balmer lines H_β, H_γ, and H_δ, from the SiII/SiIII ionization equilibrium, from UBVRI colors (kindly measured for us by M. Bessell, 1987), and from the ultraviolet flux of low-resolution IUE spectra (SWP as well as LWR range). The analysis has been carried out relative to the well-studied (galactic) B3IV star ι Herculis. As an illustration of the quality of the spectroscopic material, we show in Fig. 1 a sample part of the spectrum, and in Fig. 2 the Balmer line profiles for both Magellanic Cloud B stars. Figs. 3 and 4 exhibit the observed colors and ultraviolet (IUE) fluxes together with the calculated model fluxes for the two stars.

The model atmospheres have been calculated upon the assumption of planeparallel stratification, hydrostatic and radiative equilibrium, and LTE with Kurucz' ATLAS 6 program (1979) and his line opacity distribution functions. For the line profiles and equivalent widths a modified code by Baschek et al. (1966) has been used. The resulting element abundances, relative to the standard ι Her, are given in Table 1 for three values of the microturbulent velocity $\xi = 0, 5$, and $10\,\mathrm{km\,s^{-1}}$. Whereas for the LMC star a sufficient number of lines is available to estimate ξ to be somewhat larger than $5\,\mathrm{km\,s^{-1}}$, the data for the fainter and metal-poorer SMC star do not permit a determination of ξ. We may, however, regard $\xi = 5\,\mathrm{km\,s^{-1}}$ as a plausible value in this case. The accuracy of the relative abundances is estimated ± 0.2 in the logarithm for the LMC star, and ± 0.3 for the SMC star.

These results exemplify the present possibilities and limitations of extragalactic high-resolution spectroscopy. In this contribution, the individual abundances cannot be discussed further, we only wish to draw attention to the abundances of oxygen and iron which seem to differ from values found in some other Magellanic Cloud stars and HII regions. Details of the analysis and a discussion of the element abundances will be given elsewhere (Reitermann 1989, Reitermann et al. 1989). For information about abundances in the Magellanic Clouds, see e.g. Dufour (1984) and Pagel and Edmunds (1981).

As a further example of modern high-resolution spectroscopy of stars in near-by galaxies a few spectra of *luminous blue variables* (Hubble-Sandage variables) in the Andromeda galaxy M31 (Variable 15) and in M33 (Variable C) are shown in Figs. 5 to 9. Var 15 has been observed at V = 17.1 mag, Var C at 16.0 mag. The latter star reaches 15.2 mag at maximum, its absolute visual magnitude is $M_{V,max} \simeq M_{bol} \simeq -9.8\,\mathrm{mag}$ with the interstellar extinction being $A_v = 0.8\,\mathrm{mag}$ (cf. Humphreys et al. 1988). The spectra have been obtained by the group of the Landessternwarte Heidelberg-Königstuhl with the 3.5 m telescope at Calar Alto, Spain. The CCD spectra have a resolution of 45 Å mm^{-1} or $\Delta\lambda \simeq 1.5$ Å. Each spectrum has been exposed for one hour, yielding S/N ratios between

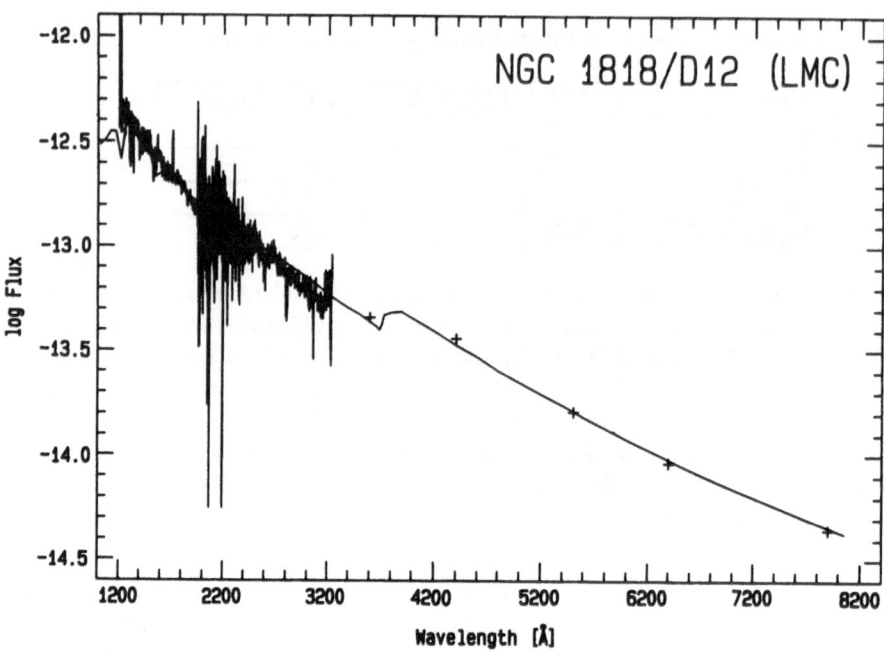

Fig. 3. Energy distribution of the B star NGC 1818/D12 in the Large Magellanic Cloud. UBVRI colors (Bessell, 1987) and low-resolution IUE spectra together with the flux calculated from a blanketed Kurucz model of $T_{eff} = 17900\,$K and $\log g = 2.85$ (Reitermann, 1989)

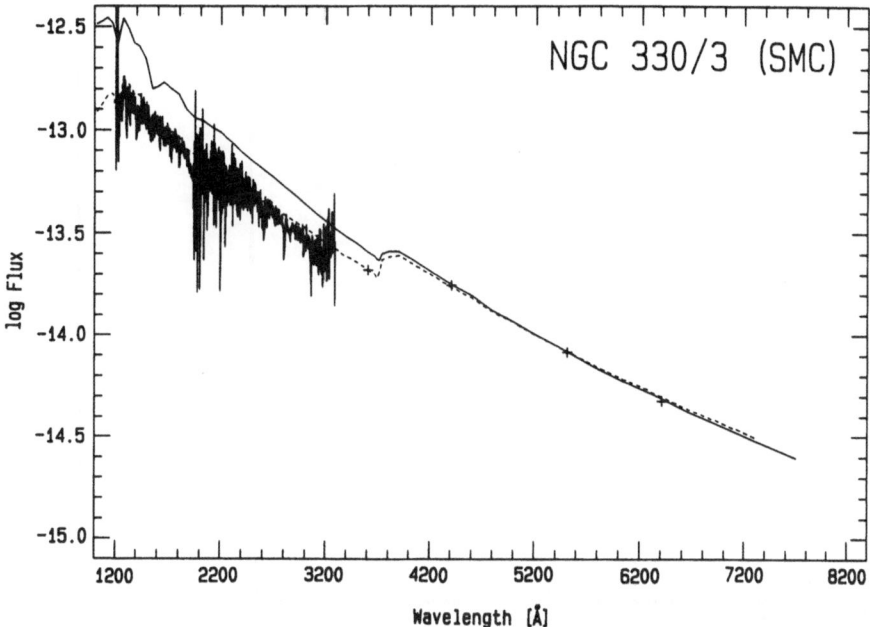

Fig. 4. Energy distribution of the B star NGC 330/3 in the Small Magellanic Cloud. UBVRI colours (Bessell, 1987) and low-resolution IUE spectra. Fluxes calculated from blanketed Kurucz models of $T_{eff} = 21000$ K, $\log g = 3.25$ (full line) and $T_{eff} = 17000$ K, $\log g = 2.8$ (dashed line); Reitermann (1989).

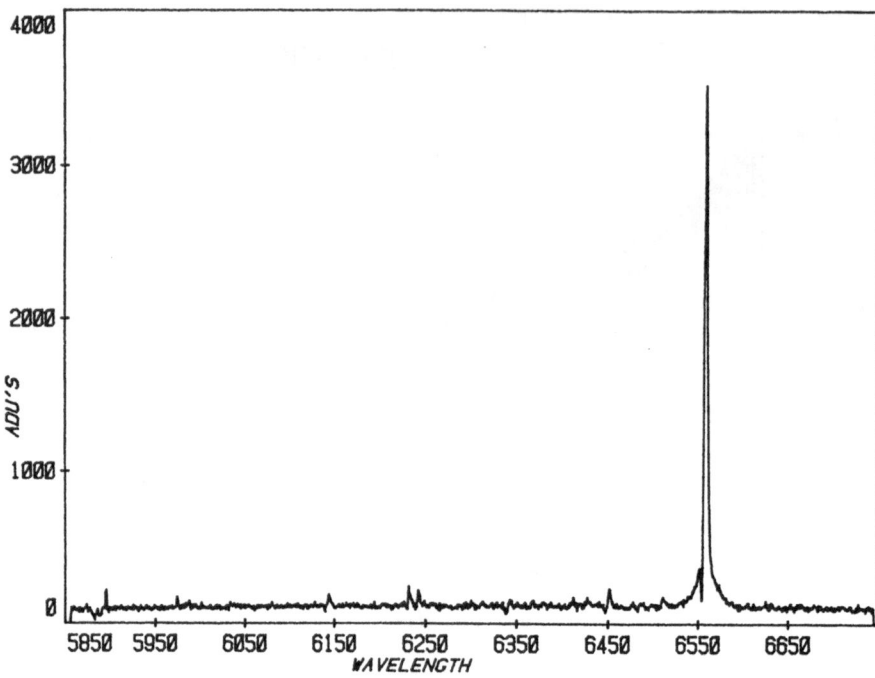

Fig. 5. Variable 15 in M31. Section of the spectrum containing H_α (Stahl and Wolf, 1989)

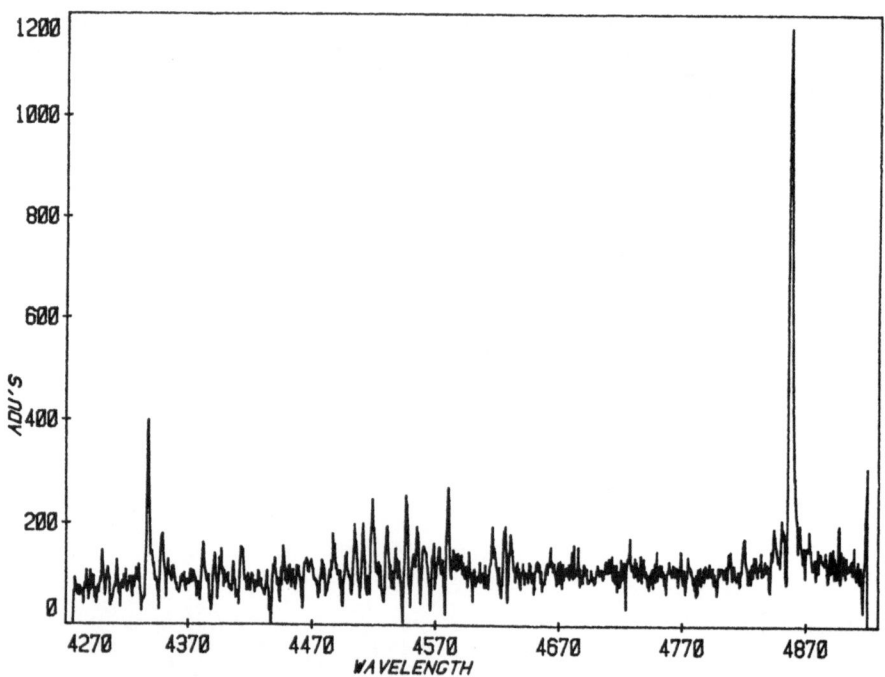

Fig. 6. Variable 15 in M31. Section of the spectrum containing H_β (Stahl and Wolf, 1989)

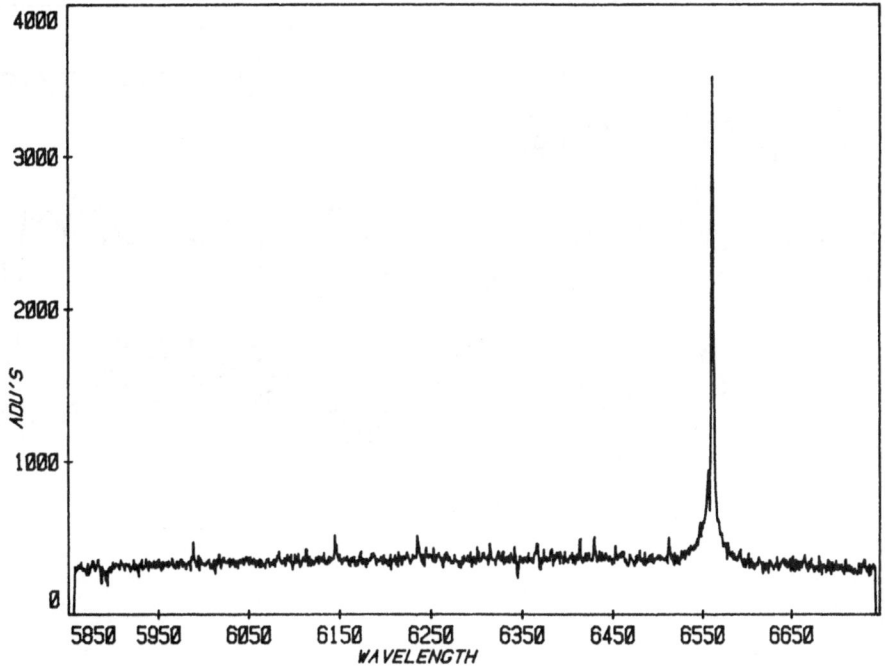

Fig. 7. Variable C in M33. Section of the spectrum containing H_α (Stahl and Wolf, 1989)

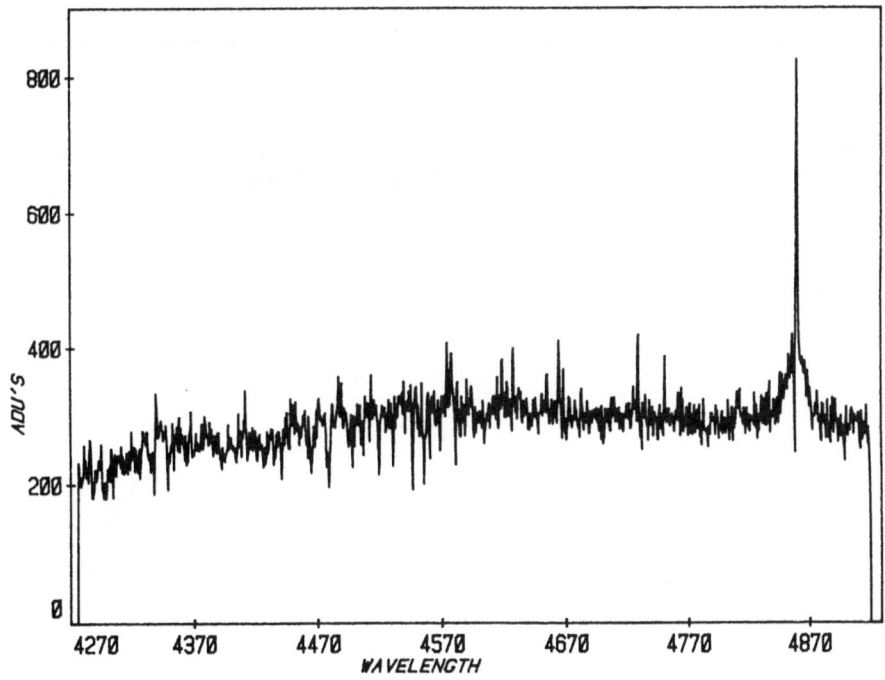

Fig. 8. Variable C in M33. Section of the spectrum containing H_β (Stahl and Wolf, 1989)

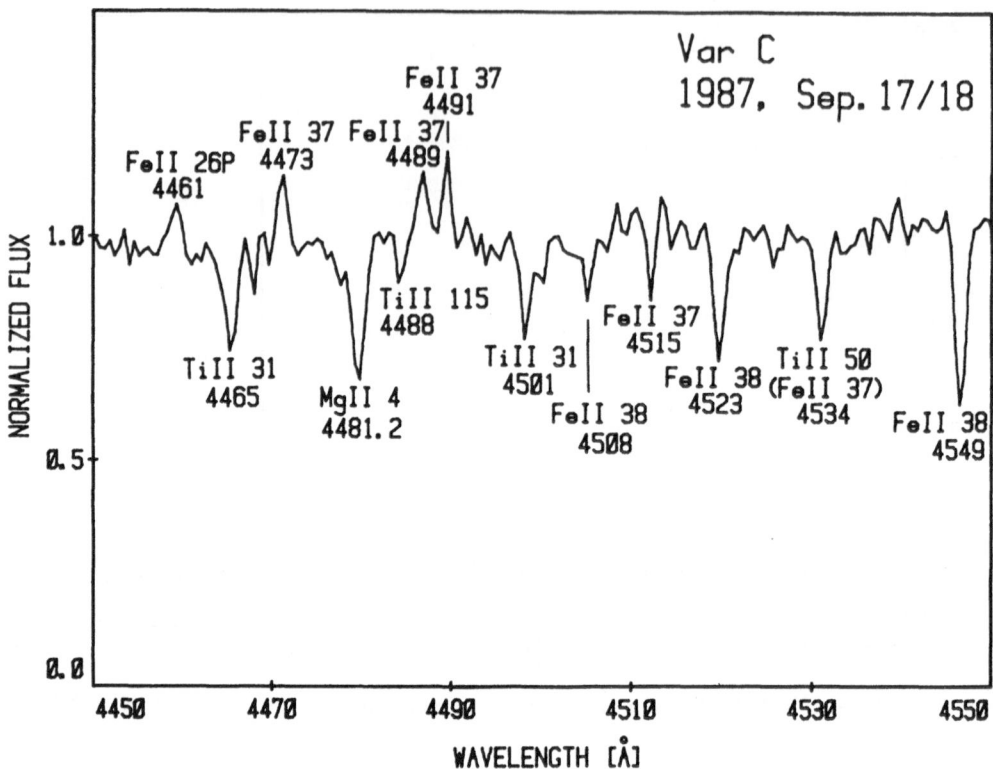

Fig. 9. Variable C in M33. Sections of the spectrum exhibiting numerous absorption and emission lines of singly ionized metals (Humphreys et al., 1988)

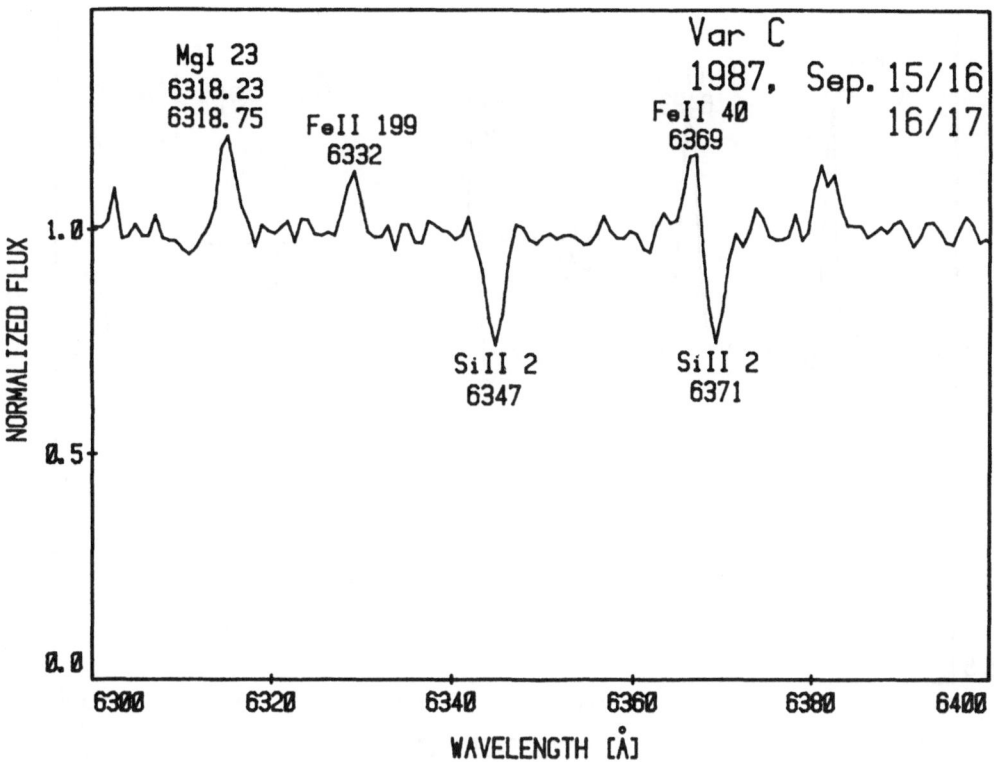

Fig. 9. Continued

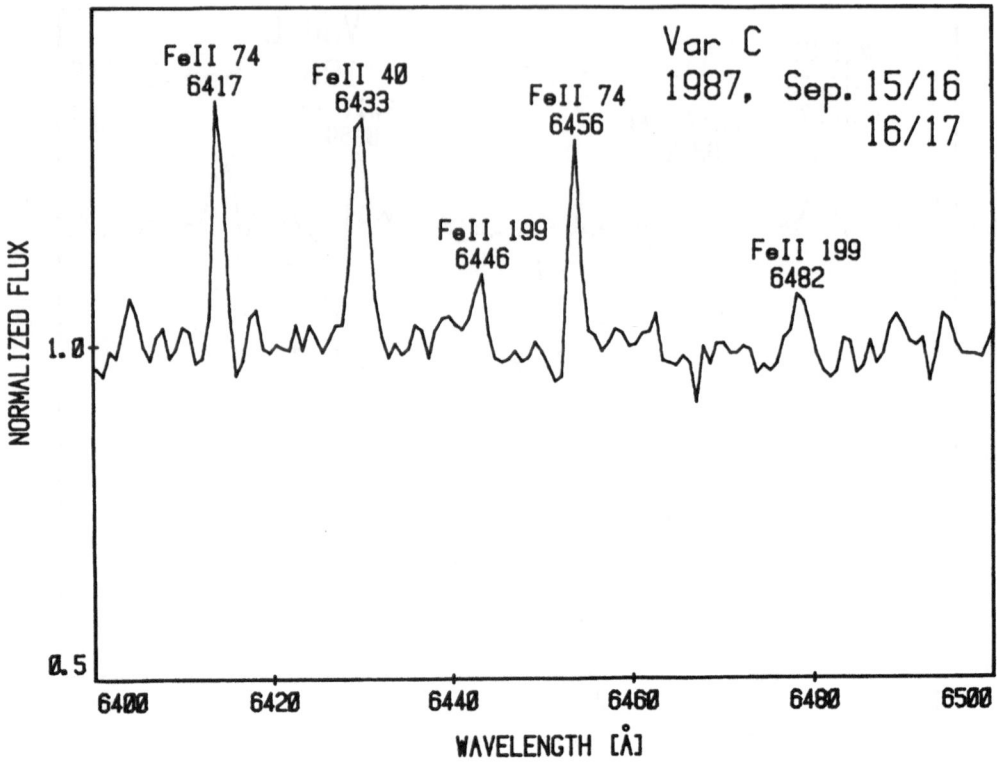

Fig. 9. Continued

30 and 50. It should be noted that the spectra also cover the *red* range. In particular the P Cygni profiles of H_α (Figs. 5 and 7) allow the determination of a mass loss rate of several 10^{-5} $M_\odot$ per year (cf. Humphreys et al. 1988). Abundance analyses of these luminous blue variables are yet to be carried out; they will have to take into account non-LTE effects.

Table 1. *Abundances of two near-main sequence B stars in blue globular clusters of the Magellanic Clouds.* The element abundances are expressed relative to the standard star ι Her, i.e. $\Delta \log\epsilon = \log\epsilon\,(\text{MC star}) - \log\epsilon(\iota\,\text{Her})$, and are given for three different microturbulent velocities ξ ; the most plausible value is $\xi = 5\,\text{km s}^{-1}$. (N) is the number of spectral lines used for the abundance determination

star	LMC NGC 1818/D12			SMC NGC 330/3		
$\xi\,[\text{km s}^{-1}]$	0	5	10(N)	0	5	10(N)
He	+0.2	+0.1	0.0(11)	0.0	−0.1	−0.1(10)
C	−0.1	−0.3	−0.5(3)	−0.8	−0.8	−0.9(3)
N	+0.6	+0.5	+0.3(16)	−0.2	−0.3	−0.4(3)
O	+0.4	+0.2	+0.1(15)	0.0	−0.1	−0.2(6)
Mg	−0.5	−0.8	−1.0(1)	−0.5	−0.8	−0.8(1)
Al	−0.4	−0.5	−0.5(4)		< −0.6	
Si	+0.1	−0.2	−0.4(4)	−0.3	−0.5	−0.5(4)
S	−0.1	−0.3	−0.3(3)		< −0.4	
Fe	−0.1	−0.3	−0.4(1)	−0.3 :	−0.4 :	−0.5 : (1)

3. POSSIBILITY OF SPECTROSCOPY OF INDIVIDUAL STARS IN GALAXIES

We now give an order-of-magnitude estimate of the limits to which spectra of individual stars can be obtained in a galaxy. Besides the star's brightness and the instrumental specifications (telescope aperture, spectral resolution) the main limiting factors are the confusion due to other stars within the aperture and the sky brightness. The latter becomes noticeable beyond about 23 mag.

The confusion problem can be evaluated by considering the number N of resolved (spatial) elements of the galaxy, the "number of beams". At a galaxian distance r, the angular resolution of the telescope corresponds to a linear dimension

$$d = 4.8\,10^{-6}\,\Phi\,r \qquad (1)$$

where Φ is expressed in seconds of arc. For simplicity we assume the galaxy to be a flat cylinder with a (linear) diamter D_G so that

$$N \leq (\frac{D_G}{d})^2 \qquad (2)$$

with the most favourable case being the galaxy seen face-on. In the following, we adopt more or less arbitrarily

$$N = \frac{1}{3} (\frac{D_G}{d})^2 \qquad (3)$$

in order to account for a typical mean inclination. Introducing adequate units we obtain with equ.(1)

$$N = 5.8 \, 10^6 \, (\frac{D_G}{20 \, \text{kpc}}) (\frac{\Phi}{1''})^{-2} (\frac{r}{1 \, \text{Mpc}})^{-2} \quad . \qquad (4)$$

Taking $\Phi = 1''$ as typical for an earth-bound telescope, and $D_G = 20$ kpc as typical for a large spiral galaxy, then at the distance of M31, $r = 0.7$ Mpc, $N \simeq 10^7$. This means that nowhere near all of the some 10^{11} stars in this galaxy can be resolved individually. On the other hand, individual OB stars with a total number of roughly 10^7 can essentially be observed individually if a suitable spectral range is chosen so that the contribution of the more numerous fainter stars of other type in the aperture is negligible. For the near-by Large Magellanic Cloud with $D_G \simeq 7$ kpc at $r \simeq 0.05$ Mpc we find $N \simeq 3 \, 10^8$, i.e. even here all stars could not be resolved.

Conversely, we can rewrite equ.(4) to obtain the maximum distance for individual spectroscopy of N very bright stars which is roughly equal to the number of resolvable elements. For a galaxy with $D_G = 20$ kpc,

$$r_{max}[\text{Mpc}] = \frac{2500}{\sqrt{N}} \quad . \qquad (5)$$

If there are about 10^4 supergiants in the galaxy, individual observation would be limited to about 25 Mpc distance, and one brilliant supernova would be observable up to about 2.5 Gpc.

With the angular resolution of $0.3''$ of the high-resolution spectrograph of the Hubble Space Telescope a factor of 3 in r_{max} or a factor of 10 in N would be gained over the seeing-dominated earth-bound observations.

However, as will be seen below, the confusion limit is not relevant for high-resolution spectroscopy of individual stars. Even for the forthcoming very large telescopes their overall sensitivity (including spectrograph and detector) for any reasonable exposure time sets a limit which is considerably sharper than that given by equ.(5).

4. THE REACH OF FUTURE VERY LARGE TELESCOPES

In the 1990s the first large ground-based telescopes with effective diameters ≥ 10 m will become available. Combined with large spectrographs these instruments will considerably extend our knowledge on element abundances in fainter stars and hence more distant galaxies. Detailed discussions of the possibilities of high-resolution, high S/N spectroscopy with very large telescopes are given e.g. by Appenzeller (1983, 1986), D'Odorico (1983), Nissen (1983), and Pagel (1986).

A very large telescope on the ground will be particularly useful for spectroscopy of "medium faint" stars, i.e. stars $\leq 23\,$mag which are brighter than the sky background so that the noise in the detector will be essentially due to photon statistics. On the other hand, for stars fainter than about 23 mag a large ground-based telescope will be inferior even to a space telescope of moderate diameter.

In the following, we give a brief overview of the parameters relevant for obtaining high-resolution spectra with photon statistics dominated noise and discuss the limitations of extragalactic spectroscopy of individual stars.

The amount of abundance information inherent in a spectrum depends on the *spectral resolution* $R = \frac{\lambda}{\Delta\lambda}$ as well as on the *S/N ratio*. For example, the minimum recognizable or measurable equivalent width is

$$W_{\lambda,min} = \eta \frac{\Delta\lambda}{S/N} = \eta \frac{\lambda}{R\,S/N} \tag{6}$$

where $\eta \simeq 1$ is required in order to detect a faint line against the noise, and $\eta \simeq 3...4$ is required to measure the line strength reliably. In the spectra of the Magellanic Cloud B stars, described in Section 2, equivalent widths could be measured with fair accuracy down to $W_{\lambda,min} \simeq 15$ mÅ, corresponding to $\eta \simeq 3$.

A desired R (or $\Delta\lambda$) and S/N has to be achieved in an *exposure time* τ with a *telescope diameter* D (and some overall efficiency Q of the telescope plus spectrograph plus detector). Finally, the limiting magnitude m of a star (above sky background) is determined by the relationship

$$10^{0.4\,m} = const.\, Q\, D^2\, \frac{1}{(S/N)^2}\Delta\lambda\,\tau \quad . \tag{7}$$

Obtaining e.g. a spectrum of a B star of $m_v = 14.2\,$mag in the Magellanic Clouds with $D = 3.6\,$m in $\tau = 2\,$h with $\Delta\lambda = 0.25$ Å and $S/N = 50$ corresponds to an overall efficiency $Q \simeq 0.2\,\%$ (cf. Nissen 1983).

As long as we aim at high-resolution spectroscopy with a good S/N ratio, the requirement of $R \simeq 2\,10^4$ and $S/N \simeq 50$ cannot be much reduced. Also the exposure time is rather limited in practice for various reasons, available telescope time, disturbance by cosmics etc. By suitable reduction procedures, e.g. by binning of the signal, one may gain perhaps a factor of 2. Essentially, however, we are left with the demand for an increase of telescope diameter. For example, the *Keck telescope* of the California Association for Research in Astronomy on Mauna Kea, Hawaii, with $D = 10\,$m will extend the limiting magnitude (equ.7) by $\Delta m = 2.2\,$mag beyond the present limit of a 3.6 m telescope. The *ESO-Very Large Telescope* in Chile with D = 16 m will gain $\Delta m = 3.2\,$mag.

Hence with the ESO-VLT early main-sequence A-type stars in the Magellanic Clouds will be accessible to high-resolution spectroscopy. Furthermore, since the difference in the distance moduli between the Clouds and M31 is about 5 mag, with some effort and the use of highly efficient spectrographs and detectors it will just be possible to obtain spectra of near-main sequence B stars with a quality comparable to that of the corresponding stars in the Magellanic Clouds (cf. Section 2).

Table 2. Elements accessible for analysis of optical spectra of B stars and their dependence on spectral resolution and distance of the parent galaxy, respectively

R	$\Delta\lambda$ [Å]	δ [Å mm^{-1}]	Δv [km s^{-1}]	$W_{\lambda,min}$ [Å]	observable elements in B stars	Δm [mag]	$\Delta(m-M)$ [mag]	galaxies
$2\,10^4$	0.25	10	15	0.015	H He C N O Mg Al Si S Fe	0	0	LMC/SMC
3000	1.5	60	90	0.090	H He C N O Mg Si	2.1		
1500	3	120	180	0.18	H He (C N O) Mg (Si)	2.8		
100	50	2000	3000	3	H	5.8	5	M31/M33
50	100		6000	6		6.5	8.5	M81
5	1000		60000	60		9.0	9.5	M51/M101

If we drop the requirement of high resolution and restrict ourselves to $R \simeq 3\,10^3$ as was used for the spectroscopy of luminous blue variables in M31 and M33 (Figs. 5 to 9), galaxies to a distance of about 5 Mpc, e.g. M81, M51, or M101, are within reach.

For spectroscopy of individual stars in even more distant galaxies there is no way but to study stars with an absolute brightness as high as possible and/or to reduce the spectral resolution and hence loose information and accuracy on the element abundances.

5. DEGRADATION OF ACCURACY AND ABUNDANCE INFORMATION DUE TO DECREASING SPECTRAL RESOLUTION

Before discussing the loss of accuracy and information about element abundances due to decreasing resolution R, we present - for convenience - various quantities equivalent to R. In Table 2, left part, are given the corresponding wavelength interval $\Delta\lambda = \lambda/R$ for a reference wavelength $\lambda = 5000$ Å, the reciprocal linear dispersion δ (adopting a "graininess" or size of detector element of 25 μm), the spread in radial velocity range $\Delta v = c\Delta\lambda/\lambda = c/R$, and the minimum equivalent width $W_{\lambda,min}$ according to equ.(6) with $\eta = 3$ and $S/N = 50$.

The range of R considered, from $2\,10^4$ down to 5, represents a steady transition from high-resolution (calibrated) spectra to narrow and wide band photometry. We note that $R = 3000$ corresponds to the refined classification used by Walborn (1970) to describe CNO anomalies in early-type stars. $R = 1200$ is equivalent to the resolution of the MK classification or, alternatively, to a rotational broading of the lines of the order of $v\,sin\,i \simeq 100\,\mathrm{km\,s^{-1}}$. With $R = 500$ Wolf-Rayet stars and carbon stars can still be discriminated by their characteristic spectral features. Finally, $R = 50$ corresponds about to the resolution of Strömgren photometry, and $R = 5$ to that of UBV photometry.

The *accuracy* of an abundance determination is discussed in some detail by R. Wehrse in this Volume. His equ.(6) shows that the error $\Delta\epsilon$ in the derived abundance ϵ can be expressed in terms of the uncertainty in the observed flux ΔF at any given wavelength and of various flux derivatives with respect to uncertainties ΔX_i in physical data (e.g. cross-sections) and model atmosphere parameters such as effective temperature and gravity:

$$\Delta\epsilon = (\frac{dF}{d\epsilon})^{-1}\left\{\Delta F + \sum_i \frac{dF}{dX_i}\Delta X_i\right\} \tag{8}$$

with $\quad \dfrac{dF}{d\epsilon} = \dfrac{\partial F}{\partial \epsilon} + \dfrac{\partial F}{\partial T}\dfrac{\partial T}{\partial \epsilon}$

While in detailed abundance studies of brighter galactic stars, the observational error $\Delta F \sim F/(S/N)$ is usually unimportant compared to the terms due to the various ΔX_i, it gains an increasing influence in the spectroscopy of stars in more and more distant galaxies. Since abundance determinations are essentially based upon the observed *equivalent widths*, their accuracy does not depend entirely on the error ΔF of the flux, but also on the spectral resolution R. As can be seen from the expression of the minimum measurable equivalent width (equ.6) S/N and R are of equal importance. Instead of discussing the $W_{\lambda,min}$ in terms of varying R for a fixed S/N ratio (Table 2) we equivalently could consider varying S/N at a given resolution R.

As an example for the degradation of information about element abundances we choose optical spectra of near-mainsequence B stars. At the high resolution of $R = 2\,10^4$ observations of Magellanic Cloud stars with $S/N \geq 50$ at a 3.6 m telescope (Reitermann et al., 1989, cf. Section 2) yield information about H, He, C, N, O, Mg, Al, Si, S, and Fe. Starting with this reference case, we show in the middle part of Table 2 the reduction of available elements with decreasing R on the basis of $W_{\lambda,min}$. For $R \leq 1000$ no elements heavier than He are accessible from optical spectra in main sequence B stars. According to equ.(7) this corresponds to an increase in magnitude or distance modulus of 5 mag. Clearly, similar considerations can be elaborated for other types of stars, too.

We conclude with a few examples of present-day observations and qualitative abundance information of the most luminous stars in galaxies. The prospects with future larger telescopes can then be judged by scaling the present data according to Table 2.

Observations of *luminous blue variables* in M31 and M33 have already been discussed in Section 2. *Novae* reach similar luminosities and could, in principle, give similar information about abundances.

Wolf-Rayet stars and carbon stars, although less luminous, exhibit marked spectral features which can be recognized even at fairly low resolution ($\Delta\lambda \geq 5$ Å). Moffat and Shara (1983, 1987) utilized interference filters of $\Delta\lambda = 87$ Å and 37 Å, respectively, to recognize the (combined) emissions of HeII, N and C around $\lambda = 4670$ Å being characteristic of *Wolf-Rayet stars*. They detected these stars which typically have $M_v \simeq M_B \simeq -4.5$ mag down to 21.5 mag in the blue. Reticon spectra with $\Delta\lambda \simeq 8$ Å obtained at the Multiple Mirror Telescope served for subsequent classification. The survey for *M supergiants* ($M_v \simeq -8.0$ mag) in M31 by Humphreys (1979) based on near-infrared spectra with $\Delta\lambda \simeq 3...5$ Å is complete to 19th magnitude. Richer and Westerlund (1983) used infrared plates with 1700 Å mm^{-1} at $\lambda = 7300$ Å (corresponding to $\Delta\lambda \simeq 50$ Å) for their survey of *carbon stars* (typically $M_I \simeq -5$ mag) in the Local Group, reaching $I \simeq 17.5$ mag. Optical IDS spectra with $\Delta\lambda = 5$ Å already exhibit a wealth of spectral details. On the other hand, even a three filter narrow-band system with $\Delta\lambda \simeq 100$ Å in the near infrared suffices to separate M stars, carbon stars, and earlier stars (Richer et al., 1984).

6. OUTLOOK

In the usual presentation of the Hertzsprung-Russell diagram, the upper part ends with the supergiants and hypergiants. We should recall, however, that *supernovae*, even a couple of weeks after maximum light, are between 5 and 9 mag brighter than the most luminous blue variables and still contribute a few per cent of the luminosity of e.g. a parent galaxy of $M_V \simeq -21$ mag. Hence they can be observed with reasonable spectral resolution up to distances of the order of 1 Gpc and are an important source of abundance information. The quantitative interpretation of supernova spectra, however, requires considerable effort and accurate abundance determination is still in its infancy. Nevertheless, qualitative results can be gained. Their light-curves which can be obtained with very low resolution, e.g. by UBV photometry, allow one to discriminate between type I and II, and hence provide information on their He/H ratio. To end, the He/H ratio in supernovae constitutes the remotest, "last bit" of stellar abundance information and is emitted from a few Gigaparsec distant galaxies.

Acknowledgments

I am very grateful to B. Wolf for helpful and stimulating discussions about many observational aspects of high-resolution spectroscopy. I wish to thank him, A. Reitermann, O. Stahl and their colleagues at the Landessternwarte Heidelberg-Königstuhl for generously putting their observational material to my disposal. Support by the Deutsche Forschungsgemeinschaft through the Sonderforschungsbereich 328 "Evolution of Galaxies" and a travel grant is gratefully acknowledged.

References

Appenzeller, I.: 1983, in Workshop on "ESO's Very Large Telescope" Eds. J.-P. Swings, K. Kjär, ESO Conf. and Workshop Proc. No. *17*, p. 7

Appenzeller, I.: 1986, in Second Workshop on "ESO's Very Large Telescope", Eds. S. D'Odorico, J.-P. Swings, ESO Conf. and Workshop Proc. No. *24*, p. 93

Baschek, B., Holweger, H., Traving, G.: 1966, Abh. Hamburger Sternwarte *8*, 26

Bessell, M.S.: 1987, unpublished

D'Odorico, S.: 1983, in Workshop on "ESO's Very Large Telescope", Eds. J.-P. Swings, K. Kjär, ESO Conf. and Workshop Proc. No. *17*, p. 37

Dufour, R.J.: 1984, in Structure and Evolution of the Magellanic Clouds, IAU Symp. No. *108* , Eds. S. van den Bergh, K.S. de Boer, D. Reidel, Dordrecht, p. 353

Humphreys, R.M.: 1979, Astrophys. J. *234*, 854

Humphreys, R.M., Leitherer, C., Stahl, O., Wolf, B., Zickgraf, F.-J.: 1988, Astron. Astrophys. *203*, 306

Kurucz, R.L.: 1979, Astrophys. J. Suppl. *40*, 1

Moffat, A.F.J., Shara, M.M.: 1983, Astrophys. J. *273*, 544

Moffat, A.F.J., Shara, M.M.: 1987, Astrophys. J. *320*, 266

Nissen, P.E.: 1983, in Workshop on "ESO's Very Large Telescope". Eds. J.-P. Swings, K. Kjär, ESO Conf. and Workshop Proc. No. *17*, p. 15

Pagel, B.E.J.: 1986, Mitt. Astron. Ges. *67*, 252

Pagel, B.E.J., Edmunds, M.G. 1981, Ann. Rev. Astron. Astrophys. *19*, 77

Reitermann, A.: 1989, PhD Thesis, University of Heidelberg

Reitermann, A., Stahl, O., Wolf, B., Baschek, B., Scholz, M.: 1989, in preparation

Richer, H.B., Crabtree, D.R., Pritchet, C.J.: 1984, Astrophys. J. *287*, 138

Richer, H.B., Westerlund, B.E.: 1983, Astrophys. J. *264*, 114

Stahl, O., Wolf, B.: 1989, unpublished

Walborn, N.R.: 1970, Astrophys. J. (Letters) *161*, L149